T0224877

Hands-on Scikit-Learn for Machine Learning Applications

Data Science Fundamentals with Python

David Paper

Apress®

Hands-on Scikit-Learn for Machine Learning Applications: Data Science Fundamentals with Python

David Paper
Logan, UT, USA

ISBN-13 (pbk): 978-1-4842-5372-4
https://doi.org/10.1007/978-1-4842-5373-1

ISBN-13 (electronic): 978-1-4842-5373-1

Managing Director, Apress Media LLC: Welmoed Spahr
Acquisitions Editor: Jonathan Gennick
Development Editor: Laura Berendson
Coordinating Editor: Jill Balzano

Cover image designed by Freepik (www.freepik.com)

Distributed to the book trade worldwide by Springer Science+Business Media New York, 233 Spring Street, 6th Floor, New York, NY 10013. Phone 1-800-SPRINGER, fax (201) 348-4505, e-mail orders-ny@springer-sbm.com, or visit www.springeronline.com. Apress Media, LLC is a California LLC and the sole member (owner) is Springer Science + Business Media Finance Inc (SSBM Finance Inc). SSBM Finance Inc is a **Delaware** corporation.

For information on translations, please e-mail rights@apress.com, or visit http://www.apress.com/rights-permissions.

Apress titles may be purchased in bulk for academic, corporate, or promotional use. eBook versions and licenses are also available for most titles. For more information, reference our Print and eBook Bulk Sales web page at http://www.apress.com/bulk-sales.

Any source code or other supplementary material referenced by the author in this book is available to readers on GitHub via the book's product page, located at www.apress.com/9781484253724. For more detailed information, please visit http://www.apress.com/source-code.

Printed on acid-free paper

For my mother, brothers, and friends.

Table of Contents

About the Author

Dr. David Paper is a professor at Utah State University in the Management Information Systems department. He is the author of two books – *Web Programming for Business: PHP Object-Oriented Programming with Oracle* and *Data Science Fundamentals for Python and MongoDB*. He has over 70 publications in refereed journals such as *Organizational Research Methods, Communications of the ACM, Information & Management, Information Resource Management Journal, Communications of the AIS, Journal of Information Technology Case and Application Research*, and *Long Range Planning*. He has also served on several editorial boards in various capacities, including associate editor. Besides growing up in family businesses, Dr. Paper has worked for Texas Instruments, DLS, Inc., and the Phoenix Small Business Administration. He has performed IS consulting work for IBM, AT&T, Octel, Utah Department of Transportation, and the Space Dynamics Laboratory. Dr. Paper's teaching and research interests include data science, machine learning, process reengineering, object-oriented programming, and change management.

About the Technical Reviewer

Jojo Moolayil is an artificial intelligence, deep learning, machine learning, and decision science professional and published author of three books: *Smarter Decisions – The Intersection of Internet of Things and Decision Science*, *Learn Keras for Deep Neural Networks*, and *Applied Supervised Learning with R*. He has worked with industry leaders on several high-impact and critical data science and machine learning projects across multiple verticals. He is currently associated with Amazon Web Services as a research scientist – AI.

Jojo was born and raised in Pune, India, and graduated from the University of Pune with a major in Information Technology Engineering. He started his career with Mu Sigma Inc., the world's largest pure-play analytics provider, and worked with the leaders of many Fortune 50 clients. He later worked with Flutura – an IoT analytics start-up – and GE, the pioneer and leader in Industrial AI.

He currently resides in Vancouver, BC. Apart from authoring books on deep learning, decision science, and IoT, Jojo has also been a technical reviewer for various books on the same subject with Apress and Packt publications. He is an active Data Science tutor and maintains a blog at http://blog.jojomoolayil.com.

- Jojo's personal web site: www.jojomoolayil.com
- Business e-mail: mail@jojomoolayil.com

Introduction

We apply the popular Scikit-Learn library to demonstrate machine learning exercises with Python code to help readers solve machine learning problems. The book is designed for those with intermediate programming skills and some experience with machine learning algorithms. We focus on application of the algorithms rather than theory. So, readers should read about the theory online or from other sources if appropriate. The reader should also be willing to spend a lot of time working through the code examples because they are pretty deep. But, the effort will pay off because the examples are intended to help the reader tackle complex problems.

The book is organized into eight chapters. Chapter 1 introduces the topic of machine learning, Anaconda, and Scikit-Learn. Chapters 2 and 3 introduce algorithmic classification. Chapter 2 classifies simple data sets and Chapter 3 classifies complex ones. Chapter 4 introduces predictive modeling with regression. Chapters 5 and 6 introduce classification tuning. Chapter 5 tunes simple data sets and Chapter 6 tunes complex ones. Chapter 7 introduces predictive modeling regression tuning. Chapter 8 puts all knowledge together to review and present findings in a holistic manner.

Download this book's example data by clicking the Download source code button found on the book's catalog page at `https://www.apress.com/us/book/9781484253724`.

CHAPTER 1

Introduction to Scikit-Learn

Scikit-Learn is a Python library that provides simple and efficient tools for implementing supervised and unsupervised machine learning algorithms. The library is accessible to everyone because it is open source and commercially usable. It is built on NumPY, SciPy, and matplolib libraries, which means it is reliable, robust, and core to the Python language.

Scikit-Learn is focused on data modeling rather than data loading, cleansing, munging or manipulating. It is also very easy to use and relatively clean of programming bugs.

Machine Learning

Machine learning is getting computers to program themselves. We use algorithms to make this happen. An *algorithm* is a set of rules used to calculate or problem solve with a computer.

Machine learning advocates create, study, and apply algorithms to improve performance on data-driven tasks. They use tools and technology to answer questions about data by training a machine how to learn.

1

© David Paper 2020
D. Paper, *Hands-on Scikit-Learn for Machine Learning Applications*,
https://doi.org/10.1007/978-1-4842-5373-1_1

The goal is to build robust algorithms that can manipulate input data to *predict* an output while continually updating outputs as new data becomes available. Any information or data sent to a computer is considered *input*. Data produced by a computer is considered *output*.

In the machine learning community, input data is referred to as the *feature set* and output data is referred to as the *target*. The feature set is also referred to as the *feature space*. Sample data is typically referred to as *training data*. Once the algorithm is trained with sample data, it can make predictions on new data. New data is typically referred to as *test data*.

Machine learning is divided into two main areas: supervised and unsupervised learning. Since machine learning typically focuses on prediction based on known properties learned from training data, our focus is on supervised learning.

Supervised learning is when the data set contains both inputs (or the feature set) and desired outputs (or targets). That is, we know the properties of the data. The goal is to make predictions. This ability to supervise algorithm training is a big part of why machine learning has become so popular.

To classify or regress new data, we must train on data with known outcomes. We classify data by organizing it into relevant categories. We regress data by finding the relationship between feature set data and target data.

With *unsupervised learning*, the data set contains only inputs but no desired outputs (or targets). The goal is to explore the data and find some structure or way to organize it. Although not the focus of the book, we will explore a few unsupervised learning scenarios.

Anaconda

You can use any Python installation, but I recommend installing Python with Anaconda for several reasons. First, it has over 15 million users. Second, Anaconda allows easy installation of the desired version of Python. Third, it preinstalls many useful libraries for machine learning including Scikit-Learn. Follow this link to see the Anaconda package lists for your operating system and Python version: *https://docs.anaconda. com/anaconda/packages/pkg-docs/*. Fourth, it includes several very popular editors including IDLE, Spyder, and Jupyter Notebooks. Fifth, Anaconda is reliable and well-maintained and removes compatibility bottlenecks.

You can easily download and install Anaconda with this link: *https://www.anaconda.com/ download/*. You can update with this link: *https://docs.anaconda.com/anaconda/install/ update-version/*. Just open Anaconda and follow instructions. I recommend updating to the current version.

Scikit-Learn

Python's Scikit-Learn is one of the most popular machine learning libraries. It is built on Python libraries NumPy, SciPy, and Matplotlib. The library is well-documented, open source, commercially usable, and a great vehicle to get started with machine learning. It is also very reliable and well-maintained, and its vast collection of algorithms can be easily incorporated into your projects. Scikit-Learn is focused on modeling data rather than loading, manipulating, visualizing, and summarizing data. For such activities, other libraries such as NumPy, pandas, Matplotlib, and seaborn are covered as encountered. The Scikit-Learn library is imported into a Python script as *sklearn*.

Data Sets

A great way to understand machine learning application is by working through *Python* data-driven code examples. We use either Scikit-Learn, UCI Machine Learning, or seaborn data sets for all examples. The Scikit-Learn *data sets* package embeds some small data sets for getting started and helpers to fetch larger data sets commonly used in the machine learning library to benchmark algorithms on data from the world at large. The UCI Machine Learning Repository maintains 468 data sets to serve the machine learning community. Seaborn provides an API on top of Matplotlib that offers simplicity when working with plot styles, color defaults, and high-level functions for common statistical plot types that facilitate visualization. It also integrates nicely with Pandas DataFrame functionality.

We chose the data sets for our examples because the machine learning community uses them for learning, exploring, benchmarking, and validating, so we can compare our results to others while learning how to apply machine learning algorithms.

Our data sets are categorized as either classification or regression data. Classification data complexity ranges from simple to relatively complex. Simple classification data sets include load_iris, load_wine, bank.csv, and load_digits. Complex classification data sets include fetch_20newsgroups, MNIST, and fetch_1fw_people. Regression data sets include tips, redwine.csv, whitewine.csv, and load_boston.

Characterize Data

Before working with algorithms, it is best to understand the data characterization. Each data set was carefully chosen to help you gain experience with the most common aspects of machine learning. We begin by describing the characteristics of each data set to better understand its composition and purpose. Data sets are organized by classification and regression data.

Classification data is further organized by complexity. That is, we begin with simple classification data sets that are not complex so that the reader can focus on the machine learning content rather than on the data. We then move onto more complex data sets.

Simple Classification Data

Classification is a machine learning technique for predicting the class upon which a dependent variable belongs. A *class* is a discrete response. In machine learning, a dependent variable is typically referred to as the *target*. A class is predicted based upon the independent variables of a data set. Independent variables are typically referred to as the *feature set* or *feature space*. Feature space is the collection of features used to characterize the data.

Simple data sets are those with a limited number of features. Such a data set is referred to as one with a *low-dimensional feature space*.

Iris Data

The first data set we characterize is load_iris, which consists of Iris flower data. Iris is a multivariate data set consisting of 50 samples from each of three species of iris (*Iris setosa, Iris virginica,* and *Iris versicolor*). Each sample contains four features, namely, length and width of sepals and petals in centimeters. Iris is a typical test case for machine learning classification. It is also one of the best known data sets in the data science literature, which means you can test your results against many other verifiable examples.

The first code example shown in Listing 1-1 loads Iris data, displays its keys, shape of the feature set and target, feature and target names, a slice from the DESCR key, and feature importance (from most to least).

Listing 1-1. Characterize the Iris data set

```
from sklearn import datasets
from sklearn.ensemble import RandomForestClassifier

if __name__ == "__main__":
    br = '\n'
    iris = datasets.load_iris()
    keys = iris.keys()
    print (keys, br)
    X = iris.data
    y = iris.target
    print ('features shape:', X.shape)
    print ('target shape:', y.shape, br)
    features = iris.feature_names
    targets = iris.target_names
    print ('feature set:')
    print (features, br)
    print ('targets:')
    print (targets, br)
    print (iris.DESCR[525:900], br)
    rnd_clf = RandomForestClassifier(random_state=0,
                                     n_estimators=100)
    rnd_clf.fit(X, y)
    rnd_name = rnd_clf.__class__.__name__
    feature_importances = rnd_clf.feature_importances_
    importance = sorted(zip(feature_importances, features),
                    reverse=True)
    print ('most important features' + ' (' + rnd_name + '):')
    [print (row) for i, row in enumerate(importance)]
```

Go ahead and execute the code from Listing 1-1. Remember that you can find the example from the book's example download. You don't need to type the example by hand. It's easier to access the example download and copy/paste.

Your output from executing Listing 1-1 should resemble the following:

```
dict_keys(['data', 'target', 'target_names', 'DESCR', 'feature_names',
'filename'])

features shape: (150, 4)
target shape: (150,)

feature set:
['sepal length (cm)', 'sepal width (cm)', 'petal length (cm)', 'petal
width (cm)']

targets:
['setosa' 'versicolor' 'virginica']
```

==============	====	====	=======	=====	====================
	Min	Max	Mean	SD	Class Correlation
==============	====	====	=======	=====	====================
sepal length:	4.3	7.9	5.84	0.83	0.7826
sepal width:	2.0	4.4	3.05	0.43	-0.4194
petal length:	1.0	6.9	3.76	1.76	0.9490 (high!)
petal width:					

```
most important features (RandomForestClassifier):
(0.4604447396171521, 'petal length (cm)')
(0.4241162651271012, 'petal width (cm)')
(0.09090795402103086, 'sepal length (cm)')
(0.024531041234715754, 'sepal width (cm)')
```

The code begins by importing datasets and RandomForestClassifier packages. RandomForestClassifier is an ensemble learning method that constructs a multitude of decision trees at training time and outputs the class that is the mode of the classes.

In this example, we are *only* using it to return feature importance. The main block begins by loading data and displaying its characteristics. Loading feature set data into variable X and target data into variable y is convention in the machine learning community.

The code concludes by training RandomForestClassifier on the pandas data, so it can return feature importance. When actually modeling data, we convert pandas data to NumPy for optimum performance. Keep in mind that the keys are available because the data set is embedded in Scikit-Learn.

Notice that we only took a small slice from DESCR, which holds a lot of information about the data set. I *always* recommend displaying at least the shape of the original data set before embarking on any machine learning experiment.

Tip   RandomForestClassifier is a powerful machine learning algorithm that not only models training data, but returns feature importance.

Wine Data

The next data set we characterize is load_wine. The load_wine data set consists of 178 data elements. Each element has thirteen features that describe three target classes. It is considered a classic in the machine learning community and offers an easy multi-classification data set.

The next code example shown in Listing 1-2 loads wine data and displays its keys, shape of the feature set and target, feature and target names, a slice from the DESCR key, and feature importance (from most to least).

Listing 1-2. Characterize load_wine

```
from sklearn.datasets import load_wine
from sklearn.ensemble import RandomForestClassifier

if __name__ == "__main__":
    br = '\n'
    data = load_wine()
    keys = data.keys()
    print (keys, br)
    X, y = data.data, data.target
    print ('features:', X.shape)
    print ('targets', y.shape, br)
    print (X[0], br)
    features = data.feature_names
```

```
targets = data.target_names
print ('feature set:')
print (features, br)
print ('targets:')
print (targets, br)
rnd_clf = RandomForestClassifier(random_state=0,
                                 n_estimators=100)
rnd_clf.fit(X, y)
rnd_name = rnd_clf.__class__.__name__
feature_importances = rnd_clf.feature_importances_
importance = sorted(zip(feature_importances, features),
                    reverse=True)
n = 6
print (n, 'most important features' + ' (' + rnd_name + '):')
[print (row) for i, row in enumerate(importance) if i < n]
```

After executing code from Listing 1-2, your output should resemble the following:

```
dict_keys(['data', 'target', 'target_names', 'DESCR', 'feature_names'])

features: (178, 13)
targets (178,)

[1.423e+01 1.710e+00 2.430e+00 1.560e+01 1.270e+02 2.800e+00 3.060e+00
 2.800e-01 2.290e+00 5.640e+00 1.040e+00 3.920e+00 1.065e+03]

feature set:
['alcohol', 'malic_acid', 'ash', 'alcalinity_of_ash', 'magnesium',
'total_phenols', 'flavanoids', 'nonflavanoid_phenols', 'proanthocyanins',
'color_intensity', 'hue', 'od280/od315_of_diluted_wines', 'proline']

targets:
['class_0' 'class_1' 'class_2']

6 most important features (RandomForestClassifier):
(0.19399882779940295, 'proline')
(0.16095401215681593, 'flavanoids')
```

```
(0.1452667364559143, 'color_intensity')
(0.11070045042456281, 'alcohol')
(0.1097465262717493, 'od280/od315_of_diluted_wines')
(0.08968972021098301, 'hue')
```

Tip   To create (instantiate) a machine learning algorithm (model), just assign it to a variable (e.g., model = algorithm()). To train based on the model, just fit it to the data (e.g., model.fit(X, y)).

The code begins by importing load_wine and RandomForestClassifier. The main block displays keys, loads data into X and y, displays the first vector from feature set X, displays shapes, and displays feature set and target information. The code concludes by training X with RandomForestClassifier, so we can display the six most important features. Notice that we display the first vector from feature set X to verify that all features are numeric.

Bank Data

The next code example shown in Listing 1-3 works with bank data. The bank.csv data set is composed of direct marketing campaigns from a Portuguese banking institution. The target is described by whether a client will subscribe (yes/no) to a term deposit (target label y). It consists of 41188 data elements with 20 features for each element. A 10% random sample of 4119 data elements is also available from this site for more computationally expensive algorithms such as svm and KNeighborsClassifier.

Listing 1-3. Characterize bank data

```
import pandas as pd

if __name__ == "__main__":
    br = '\n'
    f = 'data/bank.csv'
    bank = pd.read_csv(f)
    features = list(bank)
```

```
print (features, br)
X = bank.drop(['y'], axis=1).values
y = bank['y'].values
print (X.shape, y.shape, br)
print (bank[['job', 'education', 'age', 'housing',
              'marital', 'duration']].head())
```

After executing code from Listing 1-3, your output should resemble the following:

```
['age', 'job', 'marital', 'education', 'default', 'housing', 'loan',
'contact', 'month', 'day_of_week', 'duration', 'campaign', 'pdays',
'previous', 'poutcome', 'emp.var.rate', 'cons.price.idx', 'cons.conf.idx',
'euribor3m', 'nr.employed', 'y']
```

```
(41188, 20) (41188,)
```

	job	education	age	housing	marital	duration
0	housemaid	basic.4y	56	no	married	261
1	services	high.school	57	no	married	149
2	services	high.school	37	yes	married	226
3	admin.	basic.6y	40	no	married	151
4	services	high.school	56	no	married	307

The code example begins by importing the pandas package. The main block loads bank data from a CSV file into a Pandas DataFrame and displays the column names (or features). To retrieve column names from pandas, all we need to do is make the DataFrame a list and assign the result to a variable. Next, feature set X and target y are created. Finally, X and y shapes are displayed as well as a few choice features.

Digits Data

The final code example in this subsection is load_digits. The load_digits data set consists of 1797 8 × 8 handwritten images. Each image is represented by 64 pixels (based on an 8 × 8 matrix), which make up the feature set. Ten targets are predicted represented by digits zero to nine.

Listing 1-4 contains the code that characterizes load_digits.

Listing 1-4. Characterize load_digits

```python
import numpy as np
from sklearn.datasets import load_digits
import matplotlib.pyplot as plt

if __name__ == "__main__":
    br = '\n'
    digits = load_digits()
    print (digits.keys(), br)
    print ('2D shape of digits data:', digits.images.shape, br)
    X = digits.data
    y = digits.target
    print ('X shape (8x8 flattened to 64 pixels):', end=' ')
    print (X.shape)
    print ('y shape:', end=' ')
    print (y.shape, br)
    i = 500
    print ('vector (flattened matrix) of "feature" image:')
    print (X[i], br)
    print ('matrix (transformed vector) of a "feature" image:')
    X_i = np.array(X[i]).reshape(8, 8)
    print (X_i, br)
    print ('target:', y[i], br)
    print ('original "digits" image matrix:')
    print (digits.images[i])
    plt.figure(1, figsize=(3, 3))
    plt.title('reshaped flattened vector')
    plt.imshow(X_i, cmap='gray', interpolation='gaussian')
    plt.figure(2, figsize=(3, 3))
    plt.title('original images dataset')
    plt.imshow(digits.images[i], cmap='gray',
               interpolation='gaussian')
    plt.show()
```

After executing code from Listing 1-4, your output should resemble the following:

```
dict_keys(['data', 'target', 'target_names', 'images', 'DESCR'])

2D shape of digits data: (1797, 8, 8)

X shape (8x8 flattened to 64 pixels): (1797, 64)
y shape: (1797,)

vector (flattened matrix) of "feature" image:
[ 0.  0.  3. 10. 14.  3.  0.  0.  0.  8. 16. 11. 10. 13.  0.  0.  0.  7.
 14.  0.  1. 15.  2.  0.  0.  2. 16.  9. 16. 16.  1.  0.  0.  0. 12. 16.
 15. 15.  2.  0.  0.  0. 12. 10.  0.  8.  8.  0.  0.  0.  9. 12.  4.  7.
 12.  0.  0.  0.  2. 11. 16. 16.  9.  0.]

matrix (transformed vector) of a "feature" image:
[[ 0.  0.  3. 10. 14.  3.  0.  0.]
 [ 0.  8. 16. 11. 10. 13.  0.  0.]
 [ 0.  7. 14.  0.  1. 15.  2.  0.]
 [ 0.  2. 16.  9. 16. 16.  1.  0.]
 [ 0.  0. 12. 16. 15. 15.  2.  0.]
 [ 0.  0. 12. 10.  0.  8.  8.  0.]
 [ 0.  0.  9. 12.  4.  7. 12.  0.]
 [ 0.  0.  2. 11. 16. 16.  9.  0.]]

target: 8

original "digits" image matrix:
[[ 0.  0.  3. 10. 14.  3.  0.  0.]
 [ 0.  8. 16. 11. 10. 13.  0.  0.]
 [ 0.  7. 14.  0.  1. 15.  2.  0.]
 [ 0.  2. 16.  9. 16. 16.  1.  0.]
 [ 0.  0. 12. 16. 15. 15.  2.  0.]
 [ 0.  0. 12. 10.  0.  8.  8.  0.]
 [ 0.  0.  9. 12.  4.  7. 12.  0.]
 [ 0.  0.  2. 11. 16. 16.  9.  0.]]
```

Listing 1-4 also displays Figures 1-1 and 1-2. Figure 1-1 is a reshaped flattened vector of the 500th image in the data set. Each data element in feature set X is represented as a flattened vector of 64 pixels because Scikit-Learn *cannot* recognize an 8 × 8 image matrix, so we must reshape the 500th vector to an 8 × 8 image matrix to visualize. Figure 1-2 is the 500th image taken directly from the images data set that is available when we load the data into variable *digits*.

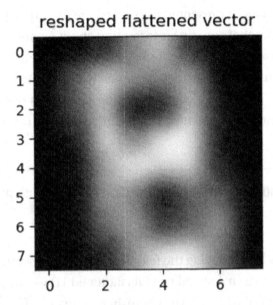

Figure 1-1. *Reshaped flattened vector of the 500th data element*

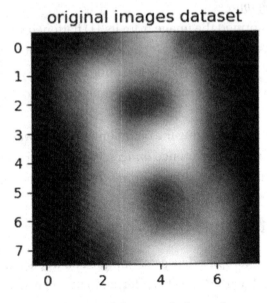

Figure 1-2. *Original image matrix of the 500th data element*

The code begins by importing numpy, load_digits, and matplotlib packages. The main block places load_digits into the *digits* variable and displays its keys: *data, target, target_names, images,* and *DESCR*. It continues by displaying the two-dimensional (2D) shape of images contained in *images*. Data in *images* are represented by 1797 8 × 8 matrices. Next, feature data (represented as vectors) are placed in X and target data in y.

A *feature vector* is one that contains information about an object's important characteristics. Data in *data* are represented by 1797 64-pixel feature vectors. A simple feature representation of an image is the raw intensity value of each pixel. So, an 8 × 8 image is represented by 64 pixels. Machine learning algorithms process feature data as vectors, so each element in *data* must be a one-dimensional (1D) vector representation of its 2D image matrix.

Tip   Feature data must be composed of vectors to work with machine learning algorithm.

The code continues by displaying the feature vector of the 500th image. Next, the 500th feature vector is transformed from its flattened 1D vector shape into a 2D image matrix and displayed with the NumPy reshape function. The code continues by displaying the target value y of the 500th image. Next, the 500th image matrix is displayed by referencing *images*.

The reason we transformed the image from its 1D flattened vector state to the 2D image matrix is that most data sets don't include an *images* object like load_data. So, to visualize and process data with machine learning algorithms, we must be able to manually *flatten* images and transform flattened images back to their original 2D matrix shape.

The code concludes by visualizing the 500th image in two ways. First, we use the flattened vector X_i. Second, we reference *images*. While machine learning algorithms require feature vectors, function imshow requires 2D image matrices to visualize.

Complex Classification Data

Now let's work with more complex data sets. Complex data sets are those with a very high number of features. Such a data set is referred to as one with a *high-dimensional feature space*.

Newsgroup Data

The first data set we characterize is fetch_20newsgroups, which consists of approximately 18000 posts on 20 topics. Data is split into train-test subsets. The split is based on messages posted before and after a specific date.

Listing 1-5 contains the code that characterizes fetch_20newsgroups.

Listing 1-5. Characterize fetch_20newsgroups

```
from sklearn.datasets import fetch_20newsgroups

if __name__ == "__main__":
    br = '\n'
    train = fetch_20newsgroups(subset='train')
    test = fetch_20newsgroups(subset='test')
    print ('data:')
    print (train.target.shape, 'shape of train data')
    print (test.target.shape, 'shape of test data', br)
    targets = test.target_names
    print (targets, br)
    categories = ['rec.autos', 'rec.motorcycles', 'sci.space',
                  'sci.med']
    train = fetch_20newsgroups(subset='train',
                               categories=categories)
    test = fetch_20newsgroups(subset='test',
                              categories=categories)
    print ('data subset:')
    print (train.target.shape, 'shape of train data')
    print (test.target.shape, 'shape of test data', br)
    targets = train.target_names
    print (targets)
```

After executing code from Listing 1-5, your output should resemble the following:

```
data:
(11314,) shape of train data
(7532,) shape of test data
```

15

```
['alt.atheism', 'comp.graphics', 'comp.os.ms-windows.misc', 'comp.sys.ibm.
pc.hardware', 'comp.sys.mac.hardware', 'comp.windows.x', 'misc.forsale',
'rec.autos', 'rec.motorcycles', 'rec.sport.baseball', 'rec.sport.hockey',
'sci.crypt', 'sci.electronics', 'sci.med', 'sci.space', 'soc.religion.
christian', 'talk.politics.guns', 'talk.politics.mideast', 'talk.politics.
misc', 'talk.religion.misc']

data subset:
(2379,) shape of train data
(1584,) shape of test data

['rec.autos', 'rec.motorcycles', 'sci.med', 'sci.space']
```

The code begins by importing fetch_20newsgroups. The main block begins by loading train and test data and displaying their shapes. Training data consists of 11314 postings, while test data consists of 7532 postings. The code continues by displaying target names and categories. Next, train and test data are created from a subset of categories. The code concludes by displaying shapes and target names of the subset.

MNIST Data

The next data set we characterize is MNIST. MNIST (Modified National Institute of Standards and Technology) is a large database of handwritten digits commonly used for training and testing in the machine learning community and other industrial image processing applications. MNIST contains 70000 examples of handwritten digit images labeled from 0 to 9 of size 28 × 28. Each target (or label) is stored as a digit value. The feature set is a matrix of 70000 28 × 28 images automatically flattened to 784 pixels each. So, each of the 70000 data elements is a vector of length 784. The target set is a vector of 70000 digit values.

Listing 1-6 contains the code that characterizes MNIST.

Listing 1-6. Characterize MNIST

```
import numpy as np
from random import randint
import matplotlib.pyplot as plt

def find_image(data, labels, d):
```

```python
    for i, row in enumerate(labels):
        if d == row:
            target = row
            X_pixels = np.array(data[i])
            return (target, X_pixels)

if __name__ == "__main__":
    br = '\n'
    X = np.load('data/X_mnist.npy')
    y = np.load('data/y_mnist.npy')
    target = np.load('data/mnist_targets.npy')
    print ('labels (targets):')
    print (target, br)
    print ('feature set shape:')
    print (X.shape, br)
    print ('target set shape:')
    print (y.shape, br)
    indx = randint(0, y.shape[0]-1)
    target = y[indx]
    X_pixels = np.array(X[indx])
    print ('the feature image consists of', len(X_pixels),
            'pixels')
    X_image = X_pixels.reshape(28, 28)
    plt.figure(1, figsize=(3, 3))
    title = 'image @ indx ' + str(indx) + ' is digit ' \
            + str(int(target))
    plt.title(title)
    plt.imshow(X_image, cmap='gray')
    digit = 7
    target, X_pixels = find_image(X, y, digit)
    X_image = X_pixels.reshape(28, 28)
    plt.figure(2, figsize=(3, 3))
    title = 'find first ' + str(int(target)) + ' in dataset'
    plt.title(title)
    plt.imshow(X_image, cmap='gray')
    plt.show()
```

After executing code from Listing 1-6, your output should resemble the following:

```
labels (targets):
[0. 1. 2. 3. 4. 5. 6. 7. 8. 9.]
```

```
feature set shape:
(70000, 784)
```

```
target set shape:
(70000,)
```

```
the feature image consists of 784 pixels
```

Listing 1-6 also displays Figures 1-3 and 1-4. Figure 1-3 is the reshaped image of digit 1 at index 6969. Figure 1-4 is the first image of digit 7 in the data set.

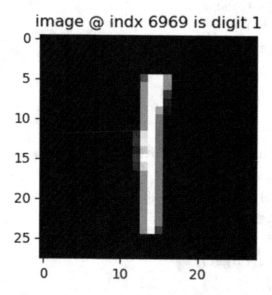

Figure 1-3. *Reshaped flattened vector of image at index 6969*

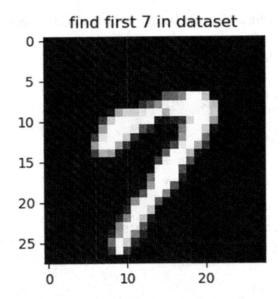

Figure 1-4. *Image of first digit 7 in the data set*

The code begins by importing randint and other requisite packages. Function find_ image locates the first occurrence of an image. The main block loads data from NumPy files into feature set X, target y, and target. Variable target holds target labels. It continues by displaying the shape of X and y. Feature set X consists of 70000 784-pixel vectors, so each image consists of 28 × 28 pixels.

Target y consists of 70000 labels. Next, a random integer between 0 and 69999 is generated, so we can display a random image from the data set. The random integer in our case is 6969. The image at index 6969 is digit 1. The size of the image is displayed to verify that it is 784 pixels. We then reshape vector 6969 to a 28 × 28 matrix, so we can visualize with function imshow. The code concludes by finding the first digit 7 and displaying it.

Faces Data

The final data set characterized in this subsection is fetch_1fw_people. The fetch_1fw_ people data set is used for classifying labeled faces. It contains 1288 face images and seven targets. Each image is represented by a 50 × 37 matrix of pixels, so the feature set has 1850 features (based on a 50 × 37 matrix). In all, the data consists of 1288 labeled faces composed of 1850 pixels each predicting seven targets.

Listing 1-7 contains the code that characterizes fetch_1fw_people.

Listing 1-7. Characterize fetch_1fw_people

```
import numpy as np
import matplotlib.pyplot as plt

if __name__ == "__main__":
    br = '\n'
    X = np.load('data/X_faces.npy')
    y = np.load('data/y_faces.npy')
    targets = np.load('data/faces_targets.npy')
    print ('shape of feature and target data:')
    print (X.shape)
    print (y.shape, br)
    print ('target faces:')
    print (targets)
    X_i = np.array(X[0]).reshape(50, 37)
    image_name = targets[y[0]]
    fig, ax = plt.subplots()
    image = ax.imshow(X_i, cmap='bone')
    plt.title(image_name)
    plt.show()
```

After executing code from Listing 1-7, your output should resemble the following:

```
shape of feature and target data:
(1288, 1850)
(1288,)

target faces:
['Ariel Sharon' 'Colin Powell' 'Donald Rumsfeld' 'George W Bush'
 'Gerhard Schroeder' 'Hugo Chavez' 'Tony Blair']
```

Listing 1-7 also displays Figure 1-5. Figure 1-5 is the reshaped image of the first data element in the data set.

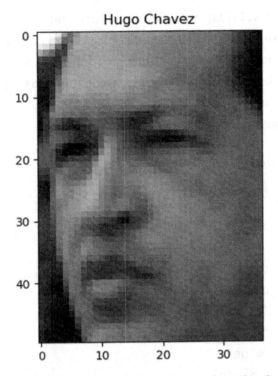

Figure 1-5. *Reshaped image of the first data element in the data set*

The code begins by importing requisite packages. The main block loads data into X, y, and targets from NumPy files. The code continues by printing shapes of X and y. X contains 1288 1850-pixel vectors and y contains 1288 target values. Target labels are then displayed. The code concludes by reshaping the first feature vector to a 50 × 37 image and displaying it with function imshow.

Regression Data

We now change gears away from classification and move into regression. *Regression* is a machine learning technique for predicting a numerical value based on the independent variables (or feature set) of a data set. That is, we are measuring the impact of the feature set on a *numerical* output. The first data set we characterize for regression is tips.

Tips Data

The tips data set is integrated with the seaborn library. It consists of food server tips in restaurants and related factors including tip, price of meal, and time of day. Specifically, features include total_bill (price of meal), tip (gratuity), sex (male or female), smoker (yes or no), day (Thursday, Friday, Saturday, or Sunday), time (day or night), and size of the party. Features are coded as follows: total_bill (US dollars), tip (US dollars), sex (0=male, 1=female), smoker (0=no, 1=yes), day (3=Thur, 4=Fri, 5= Sat, 6=Sun). Tips data is represented by 244 elements with six features predicting one target. The target being tips received from customers.

Listing 1-8 characterizes tips data.

Listing 1-8. Characterize the tips data set

```
import numpy as np, pandas as pd, seaborn as sns

if __name__ == "__main__":
    br = '\n'
    sns.set(color_codes=True)
    tips = sns.load_dataset('tips')
    print (tips.head(), br)
    X = tips.drop(['tip'], axis=1).values
    y = tips['tip'].values
    print (X.shape, y.shape)
```

After executing code from Listing 1-8, your output should resemble the following:

	total_bill	tip	sex	smoker	day	time	size
0	16.99	1.01	Female	No	Sun	Dinner	2
1	10.34	1.66	Male	No	Sun	Dinner	3
2	21.01	3.50	Male	No	Sun	Dinner	3
3	23.68	3.31	Male	No	Sun	Dinner	2
4	24.59	3.61	Female	No	Sun	Dinner	4

```
(244, 6) (244,)
```

The code begins by loading tips as a Pandas DataFrame, displaying the first five records, converting data to NumPy, and displaying the feature set and target shapes. Seaborn data is automatically loaded as a Pandas DataFrame. We couldn't get feature importance because RandomForestClassifier expects numeric data. It takes a great deal of data wrangling to get the data set into this form. We will transform categorical data to numeric in later chapters.

Red and White Wine

The next two data sets we characterize are redwine.csv and whitewine.csv. Data sets redwine.csv and whitewine.csv relate to red and white wine *quality*, respectively. Both wines are composed of variants of the Portuguese Vinho Verde wine.

The feature set consists of eleven attributes. The input attributes are based on objective tests like pH (acidity or basicity of a substance) and alcohol (percent by volume). Output quality is based on sensory data reported as the median of at least three wine expert evaluations. Each expert graded wine quality on a scale from 0 (very bad) to 10 (very excellent). The red wine data set has 1599 instances while the white wine data set has 4898.

Listing 1-9 characterizes redwine.csv.

Listing 1-9. Characterize redwine

```
import pandas as pd
from sklearn.ensemble import RandomForestRegressor

if __name__ == "__main__":
    br = '\n'
    f = 'data/redwine.csv'
    red_wine = pd.read_csv(f)
    X = red_wine.drop(['quality'], axis=1)
    y = red_wine['quality']
    print (X.shape)
    print (y.shape, br)
    features = list(X)
    rfr = RandomForestRegressor(random_state=0,
                                n_estimators=100)
    rfr_name = rfr.__class__.__name__
    rfr.fit(X, y)
    feature_importances = rfr.feature_importances_
    importance = sorted(zip(feature_importances, features),
                        reverse=True)
    n = 3
    print (n, 'most important features' + ' (' + rfr_name + '):')
    [print (row) for i, row in enumerate(importance) if i < n]
```

```
for row in importance:
    print (row)
print ()
print (red_wine[['alcohol', 'sulphates', 'volatile acidity',
               'total sulfur dioxide', 'quality']]. head())
```

After executing code from Listing 1-9, your output should resemble the following:

```
(1599, 11)
(1599,)

3 most important features (RandomForestRegressor):
(0.27432500255956216, 'alcohol')
(0.13700073893077233, 'sulphates')
(0.13053941311188708, 'volatile acidity')
(0.27432500255956216, 'alcohol')
(0.13700073893077233, 'sulphates')
(0.13053941311188708, 'volatile acidity')
(0.08068199773601588, 'total sulfur dioxide')
(0.06294612644261727, 'chlorides')
(0.057730976351602854, 'pH')
(0.055499749756166, 'residual sugar')
(0.05198192402458334, 'density')
(0.05114079873500658, 'fixed acidity')
(0.049730883807319035, 'free sulfur dioxide')
(0.04842238854446754, 'citric acid')
```

	alcohol	sulphates	volatile acidity	total sulfur dioxide	quality
0	9.4	0.56	0.70	34.0	5.0
1	9.8	0.68	0.88	67.0	5.0
2	9.8	0.65	0.76	54.0	5.0
3	9.8	0.58	0.28	60.0	6.0
4	9.4	0.56	0.70	34.0	5.0

The code example begins by loading pandas and RandomForestRegressor packages. The main block loads redwine.csv into a Pandas DataFrame. It then displays feature and target shapes. The code concludes by training pandas data with RandomForestRegressor, displaying the three most important features, and displaying the first five records from the data set. RandomForestRegressor is also an ensemble algorithm, but it is used when the target is numeric or continuous.

Tip   Always hard-code *random_state* (e.g., random_state=0) for algorithms that use this parameter to stabilize results.

The white wine example follows the *exact* same logic, but output differs in terms of data set size and feature importance.

Listing 1-10 characterizes whitewine.csv.

Listing 1-10. Characterize whitewine

```
import numpy as np, pandas as pd
from sklearn.ensemble import RandomForestRegressor

if __name__ == "__main__":
    br = '\n'
    f = 'data/whitewine.csv'
    white_wine = pd.read_csv(f)
    X = white_wine.drop(['quality'], axis=1)
    y = white_wine['quality']
    print (X.shape)
    print (y.shape, br)
    features = list(X)
    rfr = RandomForestRegressor(random_state=0,
                                n_estimators=100)
    rfr_name = rfr.__class__.__name__
    rfr.fit(X, y)
    feature_importances = rfr.feature_importances_
    importance = sorted(zip(feature_importances, features),
                        reverse=True)
    n = 3
    print (n, 'most important features' + ' (' + rfr_name + '):')
    [print (row) for i, row in enumerate(importance) if i < n]
    print ()
```

```
print (white_wine[['alcohol', 'sulphates',
                   'volatile acidity',
                   'total sulfur dioxide',
                   'quality']]. head())
```

After executing code from Listing 1-10, your output should resemble the following:

```
(4898, 11)
(4898,)

3 most important features (RandomForestRegressor):
(0.24186185906056268, 'alcohol')
(0.1251626059551235, 'volatile acidity')
(0.11524332271725685, 'free sulfur dioxide')
```

	alcohol	sulphates	volatile acidity	total sulfur dioxide	quality
0	8.8	0.45	0.27	170.0	6.0
1	9.5	0.49	0.30	132.0	6.0
2	10.1	0.44	0.28	97.0	6.0
3	9.9	0.40	0.23	186.0	6.0
4	9.9	0.40	0.23	186.0	6.0

Boston Data

The final data set we characterize is load_boston. The load_boston data set contains housing prices from various Boston locations. It consists of 506 records with 13 features and a target. Target values represent the median value of owner-occupied homes.

Listing 1-11 characterizes load_boston.

Listing 1-11. Characterize load_boston

```
from sklearn.datasets import load_boston
from sklearn.ensemble import RandomForestRegressor

if __name__ == "__main__":
    br = '\n'
    boston = load_boston()
    print (boston.keys(), br)
    print (boston.feature_names, br)
```

```
X = boston.data
y = boston.target
print ('feature shape', X.shape)
print ('target shape', y.shape, br)
keys = boston.keys()
rfr = RandomForestRegressor(random_state=0,
                            n_estimators=100)
rfr.fit(X, y)
features = boston.feature_names
feature_importances = rfr.feature_importances_
importance = sorted(zip(feature_importances, features),
                    reverse=True)
[print(row) for i, row in enumerate(importance) if i < 3]
```

After executing code from Listing 1-11, your output should resemble the following:

```
dict_keys(['data', 'target', 'feature_names', 'DESCR', 'filename'])

['CRIM' 'ZN' 'INDUS' 'CHAS' 'NOX' 'RM' 'AGE' 'DIS' 'RAD' 'TAX' 'PTRATIO'
 'B' 'LSTAT']

feature shape (506, 13)
target shape (506,)

(0.45730362625767496, 'RM')
(0.35008661885681375, 'LSTAT')
(0.06518862820215894, 'DIS')
```

The code begins by importing load_boston and RandomForestRegressor. The main block displays the keys, loads data into X and y, and displays features and data shapes. The code continues by creating RandomForestRegressor and training it with X and y so that feature importance can be displayed.

Feature Scaling

Feature scaling is standardizing feature set data. Feature scaling is important when feature set data is highly varying in magnitudes, units, and range. If such data is not scaled, some machine learning algorithms may underperform because they fail to

properly account for feature set data variance. To mitigate the problem, we standardize
the variance. Standardization rescales features to the properties of a standard normal
distribution with mean of zero ($\mu = 0$) and standard deviation of one ($\sigma = 1$). That is, we
rescale features by removing the mean and scaling to unit variance.

Scikit-Learn applies StandardScaler, which standardizes features by removing
the mean and scaling to unit variance. The code example shown in Listing 1-12
demonstrates how StandardScaler works with a machine learning algorithm.

Listing 1-12. Scaling load_digits

```
import numpy as np
from sklearn.datasets import load_digits
from sklearn.model_selection import train_test_split
from sklearn.linear_model import SGDClassifier
from sklearn.preprocessing import StandardScaler
from sklearn.metrics import accuracy_score

if __name__ == "__main__":
    br = '\n'
    digits = load_digits()
    X = digits.data
    y = digits.target
    X_train, X_test, y_train, y_test =\
            train_test_split(X, y, random_state=0)
    sgd = SGDClassifier(random_state=0, max_iter=1000,
                        tol=0.001)
    sgd.fit(X_train, y_train)
    sgd_name = sgd.__class__.__name__
    print ('<<' + sgd_name + '>>', br)
    y_pred = sgd.predict(X_test)
    accuracy = accuracy_score(y_test, y_pred)
    print ('unscaled \'test\' accuracy:', accuracy)
    scaler = StandardScaler().fit(X_train)
    X_train_std, X_test_std = scaler.transform(X_train),\
                            scaler.transform(X_test)
    sgd_std = SGDClassifier(random_state=0, max_iter=1000,
                        tol=0.001)
```

```
sgd_std.fit(X_train_std, y_train)
y_pred = sgd_std.predict(X_test_std)
accuracy = accuracy_score(y_test, y_pred)
print ('scaled \'test\' accuracy:', np.round(accuracy, 4))
```

After executing code from Listing 1-12, your output should resemble the following:

```
<<SGDClassifier>>
```

```
unscaled 'test' accuracy: 0.92
scaled 'test' accuracy: 0.9333
```

The code example begins by loading train_test_split, SGDClassifier, StandardScaler, accuracy_score, and other requisite packages. train_test_split splits vectors (data elements) from the data set into random train-test subsets. The *train* subset is used by machine learning algorithms for training while the *test* subset is used for validation.

Splitting data into train-test subsets is foundational to machine learning because models learn from train data while test data is considered *new* data that has never been seen by the model. Since test data has never been seen by the model, we can be confident that our accuracy score is valid. So, *never* use test data for training!

SGDClassifier is a classification algorithm that implements regularized linear models with stochastic gradient descent (SGD) learning. The gradient of the loss (or error) is estimated each sample at a time and the model is updated along the way with a decreasing strength schedule (or learning rate). Classification is predicting the target of given data points. Targets are also called classes, labels, or categories. Classification is covered in depth in the next two chapters.

accuracy_score is used to compute accuracy. The main block begins by placing load_digits into feature set X and target y. The code continues by splitting X and y into train-test subsets. X_train and y_train are used for training. X_test and y_test are used for validation. Next, the model is created and assigned to variable sgd. The model is then trained with X_train and y_train. Predictions are then made from X_test and assigned to y_pred. Typically, predictions are made from test data, but we can predict from train data to see how well our model is performing. The code continues by scaling train data, training the model with scaled data, and displaying accuracy. Notice scaling improved accuracy.

Dimensionality Reduction

Dimensionality (or feature) reduction is reducing the number of random variables under consideration by obtaining a set of principal variables (or components). *Principal components* are a set of values of linearly uncorrelated variables.

The premise is that data contains some features that are either redundant or irrelevant and can thereby be removed without too much information loss. Keep in mind, however, that dimensionality reduction *always* incurs some information loss.

Dimensionality reduction can simplify models, reduce training time, reduce overfitting, and avoid the curse of dimensionality. *Overfitting* is when a model trains the data too well. That is, the model understands the data perfectly but also incurs noise (or error). So, *unwanted* noise becomes part of how the model understands the data. The *curse* is prominent when working with data in high-dimensional space (that can consist of hundreds or thousands of dimensions) because it makes analyzing and organizing data very difficult. It is much easier to work with data in low-dimensional space like 2D or three-dimensional (3D) space common to the human experience.

Dimensionality reduction is useful for unsupervised learning. Unsupervised learning draws inferences from feature data without knowing their respective labeled responses (or targets). Unsupervised learning is useful for exploring hidden patterns or groupings in data.

Three common Scikit-Learn dimensionality reduction techniques are principal component analysis (PCA), LinearDiscriminantAnalysis (LDA), and Isomap. PCA and LDA are *linear* dimensionality reduction methods. Isomap is a *nonlinear* dimensionality reduction method.

The first code example shown in Listing 1-13 leverages PCA and LDA on Iris to identify clusters.

Listing 1-13. PCA and LDA Iris dimensionality reduction

```
from sklearn.datasets import load_iris
from sklearn.decomposition import PCA
from sklearn.discriminant_analysis import\
    LinearDiscriminantAnalysis
import seaborn as sns, matplotlib.pyplot as plt

if __name__ == "__main__":
    br = '\n'
    iris = load_iris()
```

```
X = iris.data
y = iris.target
pca = PCA(n_components=0.95)
X_reduced = pca.fit_transform(X)
components = pca.n_components_
model = PCA(n_components=components)
model.fit(X)
X_2D = model.transform(X)
iris_df = sns.load_dataset('iris')
iris_df['PCA1'] = X_2D[:, 0]
iris_df['PCA2'] = X_2D[:, 1]
print (iris_df[['PCA1', 'PCA2']].head(3), br)
sns.set(color_codes=True)
sns.lmplot('PCA1', 'PCA2', hue='species',
          data=iris_df, fit_reg=False)
plt.suptitle('PCA reduction')
lda = LinearDiscriminantAnalysis(n_components=2)
transform_lda = lda.fit_transform(X, y)
iris_df['LDA1'] = transform_lda[:,0]
iris_df['LDA2'] = transform_lda[:,1]
print (iris_df[['LDA1', 'LDA2']].head(3))
sns.lmplot('LDA1', 'LDA2', hue='species',
          data=iris_df, fit_reg=False)
plt.suptitle('LDA reduction')
plt.show()
```

After executing code from Listing 1-13, your output should resemble the following:

```
       PCA1      PCA2
0 -2.684126  0.319397
1 -2.714142 -0.177001
2 -2.888991 -0.144949

       LDA1      LDA2
0  8.061800  0.300421
1  7.128688 -0.786660
2  7.489828 -0.265384
```

Listing 1-13 also displays Figures 1-6 and 1-7. Figure 1-6 demonstrates how dimensionality reduction with PCA is useful for unsupervised learning visualization of clusters. Figure 1-7 demonstrates how dimensionality reduction with LDA is useful for unsupervised learning visualization of clusters.

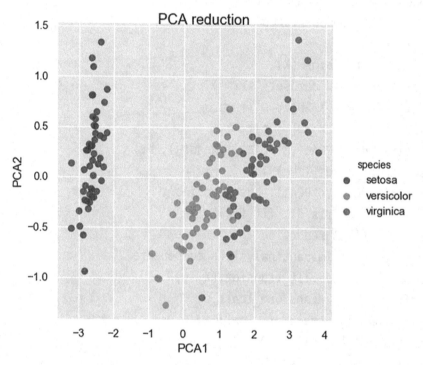

Figure 1-6. *PCA dimensionality reduction*

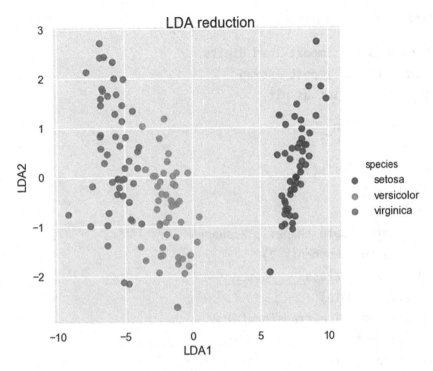

Figure 1-7. *LDA dimensionality reduction*

The code example begins by importing PCA, LinearDiscriminantAnalysis, and other requisite packages. PCA reduces the dimensionality of data consisting of many correlated variables while retaining most of its variation (or information). LinearDiscriminantAnalysis is discriminant function analysis, which is the act of distributing things into groups, classes, or categories of the same type.

The main block loads Iris into X and y. It continues by creating a PCA model with 5% information loss and transforming X into 2D space so we can determine the optimum number of principle components automatically. Next, the model is created and trained on the data. The code continues by loading Iris into a Pandas DataFrame so we can slice off the two principal components for visualization. The code continues by creating a LDA model and training it on the data.

Notice that LDA trains on both X and y data while PCA only trains on X. The code concludes by creating slices of the two principal components for visualization. Both methods do a great job of visualizing clusters for the three Iris species.

The final code example shown in Listing 1-14 uses Isomap to identify clusters on load_digits.

Listing 1-14. Isomap visualization

```python
from sklearn.datasets import load_digits
from sklearn.manifold import Isomap
import matplotlib.pyplot as plt

if __name__ == "__main__":
    br = '\n'
    digits = load_digits()
    X = digits.data
    y = digits.target
    print ('feature data shape:', X.shape)
    iso = Isomap(n_components=2)
    iso_name = iso.__class__.__name__
    iso.fit(digits.data)
    data_projected = iso.transform(X)
    print ('project data to 2D:', data_projected.shape)
    project_1, project_2 = data_projected[:, 0],\
                           data_projected[:, 1]
    plt.figure(iso_name)
    plt.scatter(project_1, project_2, c=y, edgecolor='none',
                alpha=0.5, cmap='jet')
    plt.colorbar(label='digit label', ticks=range(10))
    plt.clim(-0.5, 9.5)
    plt.show()
```

After executing code from Listing 1-14, your output should resemble the following:

```
feature data shape: (1797, 64)
project data to 2D: (1797, 2)
```

Listing 1-14 also displays Figure 1-8, which demonstrates Isomap visualization on load_digits.

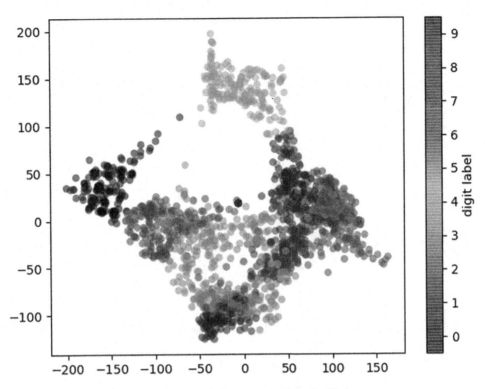

Figure 1-8. *Isomap visualization of clusters on load_digits*

The code example begins by importing Isomap and other requisite packages. The main block loads digit data into X and y and displays the shape of feature set X. The code continues by creating the Isomap model, projecting data onto 2D space, slicing principle components into variables project_1 and project_2, and visualizing. Isomap does an excellent job of identifying digit clusters 0-9.

Isomap is an excellent visualization tool for *nonlinear* data. Since load_digits data is nonlinear, Isomap worked well.

CHAPTER 2

Classification from Simple Training Sets

Classification is the problem of predicting a discrete class label. Classes are also called targets, labels, or categories. Classification is applied by training a classifier algorithm on training data to predict how new data is classified.

A machine learning classification data set consists of features (X) and targets (y) where input variables X describe known discrete output variables y. Feature data is typically referred to as the feature set (or feature space). Classification is considered supervised learning because we know the targets that correspond to the feature set.

Whew! That's a lot. So, let's look at a simple example to help you understand how classification works. Suppose we have a data set consisting of four categories of fruit, namely, "apple," "orange," "lemon," and "lime." Each data element (or row) describes one piece of fruit (the target) by its mass, width, height, and color (the features). So, an apple can be distinguished from an orange by different values of mass, width, height, and color.

In the example, the class label is the type of fruit. Each type of fruit is discrete. That is, an apple is easily distinguished from the other types of fruits. The goal is to predict type of fruit based on its mass, width, height, and color.

To train a data set, we split data into train-test subsets. Train data features are referred to as *X_train* and targets as *y_train*. Test data features are referred to as *X_test* and targets as *y_test*. We then build a classification model to train on X_train and y_train data. Once the model is trained, we can validate and predict from X_test and y_test data because the model has not seen the test data. By holding test data out of the training process, it effectively acts as *new* data.

Tip Never train on test data to keep it pure.

© David Paper 2020
D. Paper, *Hands-on Scikit-Learn for Machine Learning Applications*,
https://doi.org/10.1007/978-1-4842-5373-1_2

A typical train-test split is 70%/30%, but the ratio should be chosen based on the size of the data set. If the data set is small, a 30% test set may not contain all of the classes or enough information to properly validate. Also, the distribution of different classes in both train and test sets should be equal to the actual data set. The best way to ensure this distribution is to split train-test subsets randomly. Fortunately, Scikit-Learn's train_test_split package randomizes the split automatically, but its default train-test split is 75%/25%.

I recommend some general steps when tackling machine learning problems. First, always split data for training and validations purposes. Second, try scaling data to potentially improve performance. Third, experiment with training and test sizes. Fourth, always begin with a baseline model, simple algorithm or an algorithm based on prior experience with a data set. And, start with an algorithm's default hyperparameters. Fifth, experiment with more complex models since Scikit-Learn is efficient and allows easy model substitution. When working with big data sets, try drawing random samples to reduce computational expense. When working with high-dimensional data sets, try dimensionality reduction with PCA or LDA to reduce computational expense. Sixth, tune the best algorithms identified in earlier steps to get the best performance. Finally, experiment some more. Machine learning is very time intensive and rigorous, so be patient and don't give up.

Tip   Always begin training with an algorithm's default hyperparameters.

Simple Data Sets

We concentrate on four simple data sets to introduce machine learning classification: wine, digits, banking, and make_moons. We didn't introduce make_moons in Chapter 1 because it is contrived. That is, Scikit-Learn provides the foundation for make_moons and we construct it as we see fit.

Classifying Wine Data

The code example shown in Listing 2-1 classifies wine data.

Listing 2-1. Classify load_wine data

```python
from sklearn.datasets import load_wine
from sklearn.preprocessing import StandardScaler
from sklearn.discriminant_analysis import\
    LinearDiscriminantAnalysis as LDA
from sklearn.linear_model import SGDClassifier
from sklearn.model_selection import train_test_split
from sklearn import metrics
from random import *

if __name__ == "__main__":
    br = '\n'
    data = load_wine()
    X = data.data
    y = data.target
    X_train, X_test, y_train, y_test = train_test_split(
        X, y, test_size=0.30, random_state=0)
    lda = LDA().fit(X_train, y_train)
    print (lda, br)
    lda_name = lda.__class__.__name__
    y_pred = lda.predict(X_train)
    accuracy = metrics.accuracy_score(y_train, y_pred)
    accuracy = str(accuracy * 100) + '%'
    print (lda_name + ':')
    print ('train:', accuracy)
    y_pred_test = lda.predict(X_test)
    accuracy = metrics.accuracy_score(y_test, y_pred_test)
    accuracy = str(round(accuracy * 100, 2)) + '%'
    print ('test: ', accuracy, br)
    print('Confusion Matrix', lda_name)
    print(metrics.confusion_matrix(y_test, lda.predict(X_test)), br)
    std_scale = StandardScaler().fit(X_train)
    X_train = std_scale.transform(X_train)
    X_test = std_scale.transform(X_test)
```

```
sgd = SGDClassifier(max_iter=5, random_state=0)
print (sgd, br)
sgd.fit(X_train, y_train)
sgd_name = sgd.__class__.__name__
y_pred = sgd.predict(X_train)
y_pred_test = sgd.predict(X_test)
print (sgd_name + ':')
print('train: {:.2%}'.format(metrics.accuracy_score\(y_train, y_pred)))
print('test:  {:.2%}\n'.format(metrics.accuracy_score\(y_test, y_pred_
test)))
print('Confusion Matrix', sgd_name)
print(metrics.confusion_matrix(y_test, sgd.predict(X_test)), br)
n, ls = 100, []
for i, row in enumerate(range(n)):
    rs = randint(0, 100)
    sgd = SGDClassifier(max_iter=5, random_state=0)
    sgd.fit(X_train, y_train)
    y_pred = sgd.predict(X_test)
    accuracy = metrics.accuracy_score(y_test, y_pred)
    ls.append(accuracy)
avg = sum(ls) / len(ls)
print ('MCS (true test accuracy):', avg)
```

Go ahead and execute the code from Listing 2-1. Remember that you can find the example from the book's example download. You don't need to type the example by hand. It's easier to access the example download and copy/paste.

Your output from executing Listing 2-1 should resemble the following:

```
LinearDiscriminantAnalysis(n_components=None, priors=None, shrinkage=None,
                           solver='svd', store_covariance=False, tol=0.0001)

LinearDiscriminantAnalysis:
train: 100.0%
test:  98.15%
```

```
Confusion Matrix LinearDiscriminantAnalysis
[[19  0  0]
 [ 1 21  0]
 [ 0  0 13]]

SGDClassifier(alpha=0.0001, average=False, class_weight=None,
        early_stopping=False, epsilon=0.1, eta0=0.0,
        fit_intercept=True, l1_ratio=0.15,
        learning_rate='optimal', loss='hinge', max_iter=5,
        n_iter=None, n_iter_no_change=5, n_jobs=None,
        penalty='l2', power_t=0.5, random_state=0, shuffle=True,
        tol=None, validation_fraction=0.1, verbose=0,
        warm_start=False)

SGDClassifier:
train: 100.00%
test:  100.00%

Confusion Matrix SGDClassifier
[[19  0  0]
 [ 0 22  0]
 [ 0  0 13]]

MCS (true test accuracy): 1.0
```

The code begins by importing metrics, random, and requisite packages. The main block begins by loading data and splitting it into train-test subsets. Notice that we adjusted the test size to 30%. Next, a LinearDiscriminantAnalysis (LDA) model is created and trained on the train set. You can fiddle with test size to see if your accuracy improves. But, don't make it too big. Your model needs training data to better understand and learn.

Tip   To see a model's hyperparameters, just print the variable that holds the model after creation (e.g., print (lda)).

LDA was introduced in Chapter 1 as an unsupervised learning model for dimensionality reduction. LDA is a very interesting model in that it performs unsupervised dimensionality reduction *and* supervised classification.

Data scaling doesn't improve LDA performance, so the model trains on unscaled data. Accuracy scores are then computed on both train and test subsets. Performance accuracy is typically reported only on test data. However, it is useful to get train and test accuracy to see how well the model fits the data. In this case, the model fits the data very well because train accuracy and test accuracy are very similar. If train accuracy is well above test accuracy, the model is overfitting the data.

The code continues by displaying a confusion matrix. A *confusion matrix* describes the performance of a classification model (or classifier) on a set of test data for which the true values are known. The diagonal consisting of 19, 21, and 13 is where the model correctly classified. The model only misclassified one data element from the test set, which makes perfect sense with a test accuracy of over 98%. Next, we scale data because SGDClassifier is known to perform better with scaled data. The model is trained, and train and test accuracy are displayed along with the confusion matrix. With this model, classification was perfect.

The final part of the code is optional. It employs Monte Carlo experiments to validate performance of the SGDClassifier on the wine data. *Monte Carlo experiments* use randomness to solve deterministic (or supervised) problems. With a perfect test accuracy of 100%, we should be a bit skeptical. So, we ran 100 Monte Carlo experiments to obtain the actual test performance. As you can see, we get 100%!

Monte Carlo experiments are a fantastic method for deriving accuracy, but are incredibly computationally expensive. We were safe in this cased because the data set is small and simple. With big data sets with high-dimensional data, Monte Carlo experiments are not very practical.

LinearDiscriminantAnalysis and SGDClassifier were not chosen randomly. The algorithms were identified strategically as best performers through rigorous trial-and-error experimentation and research.

Tip   Each data set is different, so choose algorithms strategically through trial-and-error experimentation and of course research.

Classifying Digits

The first code example shown in Listing 2-2 loads the data and splits it into train-test subsets. Next, data is trained with classifiers GaussianNB, SGDClassifier, and svm. Algorithm svm is the best performer. The code then identifies and visualizes misclassifications. The code concludes by visualizing the first misclassification.

Listing 2-2. Classify load_digits data

```
from sklearn.datasets import load_digits
from sklearn.model_selection import train_test_split
from sklearn.naive_bayes import GaussianNB
from sklearn.linear_model import SGDClassifier
from sklearn.svm import SVC
from sklearn.preprocessing import StandardScaler
from sklearn.metrics import accuracy_score
from sklearn.metrics import confusion_matrix
from sklearn.metrics import classification_report
import matplotlib.pyplot as plt
import seaborn as sns

def find_misses(test, pred):
    return [i for i, row in enumerate(test) if row != pred[i]]

if __name__ == "__main__":
    br = '\n'
    digits = load_digits()
    X = digits.data
    y = digits.target
    X_train, X_test, y_train, y_test = train_test_split\
                                       (X, y, random_state=0)
    gnb = GaussianNB().fit(X_train, y_train)
    gnb_name = gnb.__class__.__name__
    y_pred = gnb.predict(X_test)
    accuracy = accuracy_score(y_test, y_pred)
    print (gnb_name + ' \'test\' accuracy:', accuracy)
    scaler = StandardScaler()
```

```
X_train_std = scaler.fit_transform(X_train)
X_test_std = scaler.fit_transform(X_test)
sgd = SGDClassifier(random_state=0, max_iter=1000, tol=0.001)
sgd_name = sgd.__class__.__name__
sgd.fit(X_train_std, y_train)
y_pred = sgd.predict(X_test_std)
accuracy = accuracy_score(y_test, y_pred)
print (sgd_name + ' \'test\' accuracy:', accuracy)
svm = SVC(gamma='auto').fit(X_train_std, y_train)
svm_name = svm.__class__.__name__
y_pred = svm.predict(X_test_std)
accuracy = accuracy_score(y_test, y_pred)
print (svm_name + ' \'test\' accuracy:', accuracy, br)
indx = find_misses(y_test, y_pred)
print ('total misclassifications (' + str(svm_name) +\ '):', len(indx), br)
print ('pred', 'actual')
misses = [(y_pred[row], y_test[row], i)
          for i, row in enumerate(indx)]
[print (row[0], '  ', row[1]) for row in misses]
img_indx = misses[0][2]
img_pred = misses[0][0]
img_act = misses[0][1]
text = str(img_pred)
print(classification_report(y_test, y_pred))
cm = confusion_matrix(y_test, y_pred)
plt.figure(1)
ax = plt.axes()
sns.heatmap(cm.T, annot=True, fmt="d",
            cmap='gist_ncar_r', ax=ax)
title = svm_name + ' confusion matrix'
ax.set_title(title)
plt.xlabel('true value')
plt.ylabel('predicted value')
test_images = X_test.reshape(-1, 8, 8)
plt.figure(2)
```

```
plt.title('1st misclassifcation')
plt.imshow(test_images[img_indx], cmap='gray', interpolation='gaussian')
plt.text(0, 0.05, text, color='r', bbox=dict(facecolor='white'))
plt.show()
```

After executing code from Listing 2-2, your output should resemble the following:

```
GaussianNB 'test' accuracy: 0.8333333333333334
SGDClassifier 'test' accuracy: 0.9377777777777778
SVC 'test' accuracy: 0.9822222222222222

total misclassifications (SVC): 8

pred actual
7    2
1    8
7    9
9    5
4    7
4    3
2    8
4    1
```

	precision	recall	f1-score	support
0	1.00	1.00	1.00	37
1	0.98	0.98	0.98	43
2	0.98	0.98	0.98	44
3	1.00	0.98	0.99	45
4	0.93	1.00	0.96	38
5	1.00	0.98	0.99	48
6	1.00	1.00	1.00	52
7	0.96	0.98	0.97	48
8	1.00	0.96	0.98	48
9	0.98	0.98	0.98	47
micro avg	0.98	0.98	0.98	450
macro avg	0.98	0.98	0.98	450
weighted avg	0.98	0.98	0.98	450

Listing 2-2 also displays Figures 2-1 and 2-2. Figure 2-1 displays the confusion matrix for the best performing algorithm, which is svm. You see *SVC* displayed because we are implementing the SVC implementation of the svm algorithm. SVC implementation utilizes C-support vector classification, which is represented as svm.SVC in Scikit-Learn. Figure 2-2 displays the first misclassification from the prediction set, which is digit 2 misclassified as digit 7. If we look at the confusion matrix, we can see this misclassification at the intersection of predicted value row for digit 7 and true value column for digit 2. So, the true value (digit 2) was incorrectly predicted (or misclassified) as digit 7.

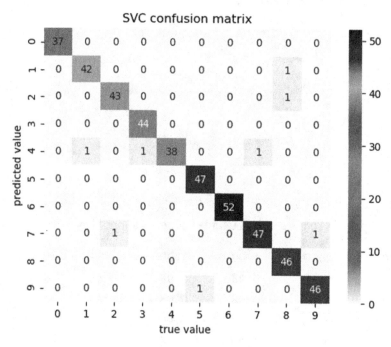

Figure 2-1. *Confusion matrix for the svm.SVC algorithm*

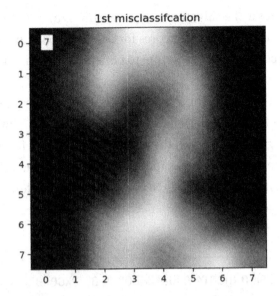

Figure 2-2. *First misclassification from the prediction set*

The code example begins by importing GaussianNB, confusion_matrix, and classification_report as well as other requisite packages. GaussianNB is an excellent baseline algorithm because it is fast, performs well with many classification problems, and has few hyperparameters to tune.

Tip If you have no experience with a classification data set, GaussianNB is a great place to start because it is simple, fast, easy to understand, and has few hyperparameters to tune.

Function find_misses returns a list of misclassified digits. The main block loads data, splits it into train-test subsets, and trains with GaussianNB, SGDClassifier, and svm.

GaussianNB is a probabilistic classifier based on applying Bayes' theorem with strong independence assumptions between features. SGDClassifier is a classifier that implements a plain stochastic gradient descent learning routine that supports different loss functions and penalties for classification. Support vector machine (svm) builds a model that assigns new examples to one category or the other.

The code then displays test accuracy for all three models. Since svm.SVC scores the highest, we use it to identify misclassifications. Total misclassifications are then displayed. The code continues by displaying how each misclassification was rendered. So, the first misclassified digit was 2 and it was misclassified as 7.

Next, the code creates a classification report (for the svm.SVC algorithm) that presents precision, recall, and f1_score scores for each digit. Accuracy is a great way to report a score, but f1_score, especially, is one to consider including because it is the most conservative.

The code concludes by displaying a svm.SVC confusion matrix and the first misclassification where 2 was misclassified as 7. The figure is also interesting because it presents the actual image of digit 2 with the way it was classified in red as digit 7.

Once a great performing algorithm is identified for a data set, I highly recommend creating a confusion matrix visualization. Not only it easy to understand how well an algorithm classified targets, it allows deeper scrutiny of where the algorithm didn't perform as expected.

Tip   Creating a confusion matrix visualization is an excellent way to get a sense of how an algorithm performs.

Although we obtained a high accuracy score from svm in the previous example, Scikit-Learn allows us to substitute classifiers very easily. So, the next code example goes a bit crazy by training wine data with six additional classifiers. Keep in mind that the data set is small and simple. With larger and more complex data, substituting classifiers can be computationally expensive.

The next code example shown in Listing 2-3 classifies wine data with several Scikit-Learn algorithms to identify promising ones for improved performance.

Listing 2-3. Classifying load_digits with various algorithms

```
import humanfriendly as hf
import time
from sklearn.datasets import load_digits
from sklearn.model_selection import train_test_split
from sklearn.linear_model import LogisticRegression,\
     LogisticRegressionCV
from sklearn.neighbors import KNeighborsClassifier
from sklearn.ensemble import RandomForestClassifier,\
     ExtraTreesClassifier, GradientBoostingClassifier
from sklearn.preprocessing import StandardScaler
```

```python
from sklearn.metrics import accuracy_score
from sklearn.metrics import f1_score

def get_scores(model, Xtest, ytest, avg):
    y_pred = model.predict(Xtest)
    accuracy = accuracy_score(ytest, y_pred)
    f1 = f1_score(ytest, y_pred, average=avg)
    return (accuracy, f1)

def get_time(time):
    return hf.format_timespan(time, detailed=True)

if __name__ == "__main__":
    br = '\n'
    digits = load_digits()
    X = digits.data
    y = digits.target
    X_train, X_test, y_train, y_test = train_test_split\
                                        (X, y, random_state=0)
    scaler = StandardScaler()
    X_train_std = scaler.fit_transform(X_train)
    X_test_std = scaler.fit_transform(X_test)

    lr = LogisticRegression(random_state=0, solver='lbfgs',
                            multi_class='auto', max_iter=4000)
    lr.fit(X_train_std, y_train)
    lr_name = lr.__class__.__name__
    acc, f1 = get_scores(lr, X_test_std, y_test, 'micro')
    print (lr_name + ' scaled \'test\':')
    print ('accuracy:', acc, ', f1_score:', f1, br)
    softmax = LogisticRegression(multi_class="multinomial",
                                 solver="lbfgs", max_iter=4000,
                                 C=10, random_state=0)
    softmax.fit(X_train_std, y_train)
    acc, f1 = get_scores(softmax, X_test_std, y_test, 'micro')
    print (lr_name + ' (softmax) scaled \'test\':')
    print ('accuracy:', acc, ', f1_score:', f1, br)
```

```python
rf = RandomForestClassifier(random_state=0, n_estimators=100)
rf.fit(X_train_std, y_train)
rf_name = rf.__class__.__name__
acc, f1 = get_scores(rf, X_test_std, y_test, 'micro')
print (rf_name + ' \'test\':')
print ('accuracy:', acc, ', f1_score:', f1, br)
et = ExtraTreesClassifier(random_state=0, n_estimators=100)
et.fit(X_train, y_train)
et_name = et.__class__.__name__
acc, f1 = get_scores(et, X_test, y_test, 'micro')
print (et_name + ' \'test\':')
print ('accuracy:', acc, ', f1_score:', f1, br)
gboost_clf = GradientBoostingClassifier(random_state=0)
gb_name = gboost_clf.__class__.__name__
gboost_clf.fit(X_train, y_train)
acc, f1 = get_scores(gboost_clf, X_test, y_test, 'micro')
print (gb_name + ' \'test\':')
print ('accuracy:', acc, ', f1_score:', f1, br)
knn_clf = KNeighborsClassifier().fit(X_train, y_train)
knn_name = knn_clf.__class__.__name__
acc, f1 = get_scores(knn_clf, X_test, y_test, 'micro')
print (knn_name + ' \'test\':')
print ('accuracy:', acc, ', f1_score:', f1, br)
start = time.perf_counter()
lr_cv = LogisticRegressionCV(random_state=0, cv=5, multi_class='auto',
                             max_iter=4000)
lr_cv_name = lr_cv.__class__.__name__
lr_cv.fit(X, y)
end = time.perf_counter()
elapsed_ls = end - start
timer = get_time(elapsed_ls)
print (lr_cv_name + ' timer:', timer)
acc, f1 = get_scores(lr_cv, X_test, y_test, 'micro')
print (lr_cv_name + ' \'test\':')
print ('accuracy:', acc, ', f1_score:', f1)
```

After executing code from Listing 2-3, your output should resemble the following:

```
LogisticRegression scaled 'test':
accuracy: 0.9733333333333334 , f1_score: 0.9733333333333334

LogisticRegression (softmax) scaled 'test':
accuracy: 0.9644444444444444 , f1_score: 0.9644444444444444

RandomForestClassifier 'test':
accuracy: 0.9755555555555555 , f1_score: 0.9755555555555555

ExtraTreesClassifier 'test':
accuracy: 0.9822222222222222 , f1_score: 0.9822222222222222

GradientBoostingClassifier 'test':
accuracy: 0.9622222222222222 , f1_score: 0.9622222222222222

KNeighborsClassifier 'test':
accuracy: 0.98 , f1_score: 0.98

LogisticRegressionCV timer: 49 seconds and 38.45 milliseconds
LogisticRegressionCV 'test':
accuracy: 0.9822222222222222 , f1_score: 0.9822222222222222
```

The code begins by importing humanfriendly, time, LogisticRegression, LogisticRegressionCV, KNeighborsClassifier, GradientBoostingClassifier, and ExtraTreesClassifier as well as other requisite packages. Function get_scores returns accuracy and f1_score. Function get_time returns elapsed time. It facilitates finding how long it takes an algorithm to train a data set.

LogisticRegression is a classification algorithm traditionally limited to only two-class classification problems. Softmax (multinomial logistic regression) classification uses logistic regression for multiclass classification. RandomForestClassifier is an ensemble learning method that constructs a multitude of decision trees at training time and outputs the class that is the mode of the classes. ExtraTreesClassifier implements a meta estimator that fits a number of randomized decision trees (or extra trees) on various subsamples of the data set and uses averaging to improve predictive accuracy and control overfitting. GradientBoostingClassifier produces a prediction model in the form of weak prediction models (typically decision trees). KNeighborsClassifier implements the k-nearest neighbors' vote where input consists of the k closest training examples in the feature space. *Feature space* refers to the n-dimensions where your features exist. LogisticRegression uses a logistic function to model data. LogisticRegressionCV uses logistic regression to implement cross-validation estimation.

Cross-validation (CV) divides data into *n* number of subsets and iterates *n* times. Through each iteration, one of the *n* subsets is held out as the test set while the rest are used for training. Every iteration uses a different subset. So, accuracy and error are averaged over all *n* trials. The resultant accuracy is very good, but CV can be computationally expensive.

GradientBoostingClassifier and ExtraTreesClassifier are ensemble methods similar to RandomForestClassifier in that they fit (or train) a number of decision trees on the data and average results to improve predictive accuracy.

Tip   You may have to install the humanfriendly package since it isn't installed automatically by Anaconda. Open a new Anaconda prompt and install as shown in Listing 2-4.

Listing 2-4. Install a new package

```
pip install humanfriendly
```

The main block begins by loading and splitting data into train-test subsets. Each of the algorithms train the data and scores are displayed. Notice that over 46 seconds are consumed by LogisticRegressionCV. Although all of the algorithms performed admirably, we still couldn't beat 98.22% accuracy.

Classifying Bank Data

The first code example shown in Listing 2-5 loads bank data from a CSV file. Next, the *education* feature is engineered to make it more presentable.

Feature engineering is creating features (based on domain knowledge of the data) that make machine learning algorithms work. Although feature engineering is fundamental to machine learning application, it is both difficult and expensive.

The code then transforms categorical features to numerical to enable algorithm training. This transformation is typically referred to as encoding.

Tip   Machine learning algorithms only operate on numerical data.

Finally, the five most important features are displayed along with data features and class counts. The feature set and targets are saved in NumPy files.

Listing 2-5. Engineering and wrangling bank data

```python
import numpy as np, pandas as pd
from sklearn.ensemble import RandomForestClassifier

if __name__ == "__main__":
    br = '\n'
    f = 'data/bank.csv'
    data = pd.read_csv(f)
    print ('original "education" categories:')
    print (data.education.unique(), br)
    data['education'] = np.where(data['education'] == 'basic.9y',
                                 'basic', data['education'])
    data['education'] = np.where(data['education'] == 'basic.6y',
                                 'basic', data['education'])
    data['education'] = np.where(data['education'] == 'basic.4y',
                                 'basic', data['education'])
    data['education'] = np.where(data['education'] == 'high.school',
                                 'high_school', data.education)
    data['education'] = np.where(data['education'] == 'professional.course',
                                 'professional', data['education'])
    data['education'] = np.where(data['education'] == 'university.degree',
                                 'university', data['education'])
    print ('engineered "education" categories:')
    print (data.education.unique(), br)
    print ('target value counts:')
    print (data.y.value_counts(), br)
    data_X = data.loc[:, data.columns != 'y']
    cat_vars = ['job', 'marital', 'education', 'default', 'housing',
                'loan', 'contact', 'month', 'day_of_week', 'poutcome']
    data_new = pd.get_dummies(data_X, columns=cat_vars)
```

```
X = data_new.values
y = data.y.values
attributes = list(data_X)
rf = RandomForestClassifier(random_state=0, n_estimators=100)
rf.fit(X, y)
rf_name = rf.__class__.__name__
feature_importances = rf.feature_importances_
importance = sorted(zip(feature_importances, attributes), reverse=True)
n = 5
print (n, 'most important features' + ' (' + rf_name + '):')
[print (row) for i, row in enumerate(importance) if i < n]
print ()
features_file = 'data/features'
np.save(features_file, attributes)
features = np.load('data/features.npy')
print ('features:')
print (features, br)
y_file = 'data/y'
X_file = 'data/X'
np.save(y_file, y)
np.save(X_file, X)
d = {}
dvc = data.y.value_counts()
d['no'], d['yes'] = dvc['no'], dvc['yes']
dvc_file = 'data/value_counts'
np.save(dvc_file, d)
d = np.load('data/value_counts.npy')
print ('class counts:', d)
```

After executing code from Listing 2-5, your output should resemble the following:

```
original "education" categories:
['basic.4y' 'high.school' 'basic.6y' 'basic.9y' 'professional.course'
 'unknown' 'university.degree' 'illiterate']

engineered "education" categories:
['basic' 'high_school' 'professional' 'unknown' 'university' 'illiterate']
```

```
target value counts:
no      36548
yes      4640
Name: y, dtype: int64
```

```
5 most important features (RandomForestClassifier):
(0.28697175347986037, 'job')
(0.08761238456151103, 'month')
(0.0797624194551633, 'age')
(0.05492109153356108, 'day_of_week')
(0.04027613029914145, 'marital')
```

```
features:
['age' 'job' 'marital' 'education' 'default' 'housing' 'loan' 'contact'
 'month' 'day_of_week' 'duration' 'campaign' 'pdays' 'previous' 'poutcome'
 'emp.var.rate' 'cons.price.idx' 'cons.conf.idx' 'euribor3m' 'nr.employed']
```

```
class counts: {'no': 36548, 'yes': 4640}
```

The code example begins by importing requisite packages. The main block reads the data and displays the original values from the *education* feature. The code continues by feature engineering the feature and displays the new values. Notice how difficult it is to feature engineer just a single feature. Next, categorical features are encoded by the pandas *get_dummies* function to one hot encoding (OHE) vectors. Scikit-Learn expects feature data to be numeric, which is why we need to encode them.

OHE vectors are also called dummy variables. OHE is a good choice since it is one of the most common methods for dealing with categorical data in machine learning. OHE takes each category value and turns it into a binary vector of size i (where i is the number of values in category i) and makes all columns equal to zero except the category column. For example, marital status is either "married," "single," or "divorced" in our data set. If someone is married, OHE encodes a [1 0 0] vector. If single, OHE encodes a [0 1 0] vector. Finally, if divorced, OHE encodes a [0 0 1] vector. Simply, the 1 bit is hot to indicate the category that fits the data element.

The code then creates feature set X and target y from the transformed data set. Feature importance is displayed with the help of RandomForestClassifier. Next, X and y are saved in NumPy files. Finally, class counts are created, saved, and displayed. It is useful to view class counts to see the balance between targets.

Notice that we have more *no* values than *yes* values. So, the data set is a bit imbalanced. This occurrence is commonly referred to as imbalanced class distribution, which is when the number of observations belonging to one class is significantly lower than those belonging to other classes. In our case, the balance between yes and no is about 12.6%. So, we don't have a major problem. A rate (or event rate) less than 5% is a problem because machine learning algorithms can produce unsatisfactory classification when this happens.

Now that bank data is prepared, we can run experiments to identify high-performing classification algorithms as demonstrated in the next code example, which is shown in Listing 2-6. Keep in mind that many hours of experimentation led to the choice of algorithms for this example.

Listing 2-6. Classifying bank data

```
import numpy as np, pandas as pd, random
from sklearn.model_selection import train_test_split
from sklearn.preprocessing import StandardScaler
from sklearn.ensemble import RandomForestClassifier,\
    ExtraTreesClassifier
from sklearn.svm import SVC
from sklearn.neighbors import KNeighborsClassifier
from sklearn.metrics import f1_score
from sklearn.metrics import confusion_matrix
import matplotlib.pyplot as plt
import seaborn as sns

def get_scores(model, xtrain, ytrain, xtest, ytest, scoring):
    ypred = model.predict(xtest)
    train = model.score(xtrain, ytrain)
    test = model.score(xtest, y_test)
    f1 = f1_score(ytest, ypred, average=scoring)
    return (train, test, f1)

def prep_data(data, target):
    d = [data[i] for i, _ in enumerate(data)]
    t = [target[i] for i, _ in enumerate(target)]
    return list(zip(d, t))
```

```python
def create_sample(d, n, replace='yes'):
    if replace == 'yes': s = random.sample(d, n)
    else: s = [random.choice(d) for i, _ in enumerate(d) if i < n]
    Xs = [row[0] for i, row in enumerate(s)]
    ys = [row[1] for i, row in enumerate(s)]
    return np.array(Xs), np.array(ys)

if __name__ == "__main__":
    br = '\n'
    X = np.load('data/X.npy')
    y = np.load('data/y.npy')
    print ('full data set shape for X and y:')
    print (X.shape, y.shape, br)
    X_train, X_test, y_train, y_test = train_test_split\
                                        (X, y, random_state=0)
    et = ExtraTreesClassifier(random_state=0, n_estimators=100)
    et.fit(X_train, y_train)
    et_scores = get_scores(et, X_train, y_train, X_test, y_test, 'micro')
    print (et.__class__.__name__ + '(train, test, f1_score):')
    print (et_scores, br)
    rf = RandomForestClassifier(random_state=0, n_estimators=100)
    rf.fit(X_train, y_train)
    rf_scores = get_scores(rf, X_train, y_train, X_test, y_test, 'micro')
    print (rf.__class__.__name__ + '(train, test, f1_score):')
    print (rf_scores, br)
    sample_size = 4000
    data = prep_data(X, y)
    Xs, ys = create_sample(data, sample_size, replace='no')
    print ('sample data set shape for X and y:')
    print (Xs.shape, ys.shape, br)
    X_train, X_test, y_train, y_test = train_test_split\
                                        (Xs, ys, random_state=0)
    scaler = StandardScaler().fit(X_train)
    X_train_std, X_test_std = scaler.transform(X_train),\
                            scaler.transform(X_test)
    knn = KNeighborsClassifier().fit(X_train, y_train)
```

```python
knn_scores = get_scores(knn, X_train, y_train, X_test, y_test, 'micro')
print (knn.__class__.__name__ + '(train, test, f1_score):')
print (knn_scores, br)
svm = SVC(random_state=0, gamma='scale')
svm.fit(X_train_std, y_train)
svm_scores = get_scores(svm, X_train_std, y_train, X_test_std, y_test,
                        'micro')
print (svm.__class__.__name__ + '(train, test, f1_score):')
print (svm_scores, br)
knn_name, svm_name = knn.__class__.__name__,\
                     svm.__class__.__name__
y_pred_knn = knn.predict(X_test)
cm_knn = confusion_matrix(y_test, y_pred_knn)
cm_knn_T = cm_knn.T
y_pred_svm = svm.predict(X_test_std)
cm_svm = confusion_matrix(y_test, y_pred_svm)
cm_svm_T = cm_svm.T
plt.figure(knn.__class__.__name__)
ax = plt.axes()
sns.heatmap(cm_knn_T, annot=True, fmt="d", cmap='gist_ncar_r', cbar=False)
ax.set_title(str(knn_name) + ' confusion matrix')
plt.xlabel('true label')
plt.ylabel('predicted label')
plt.figure(str(svm_name) + ' confusion matrix' )
ax = plt.axes()
sns.heatmap(cm_svm_T, annot=True, fmt="d", cmap='gist_ncar_r', cbar=False)
ax.set_title(svm_name)
plt.xlabel('true label')
plt.ylabel('predicted label')
cnt_no, cnt_yes = 0, 0
for i, row in enumerate(y_test):
    if row == 'no': cnt_no += 1
    elif row == 'yes': cnt_yes += 1
cnt_no, cnt_yes = str(cnt_no), str(cnt_yes)
print ('true =>', 'no: ' + cnt_no + ', yes: ' + cnt_yes, br)
```

```
p_no, p_nox = cm_knn_T[0][0], cm_knn_T[0][1]
p_yes, p_yesx = cm_knn_T[1][1], cm_knn_T[1][0]
print ('knn classification report:')
print ('predict \'no\':', p_no, '(' +\str(p_nox) + ' misclassifed)')
print ('predict \'yes\':', p_yes, '(' +\str(p_yesx) + ' misclassifed)', br)
p_no, p_nox = cm_svm_T[0][0], cm_svm_T[0][1]
p_yes, p_yesx = cm_svm_T[1][1], cm_svm_T[1][0]
print ('svm classification report:')
print ('predict \'no\':', p_no, '(' +\str(p_nox) + ' misclassifed)')
print ('predict \'yes\':', p_yes, '(' +\str(p_yesx) + ' misclassifed)')
plt.show()
```

After executing code from Listing 2-6, your output should resemble the following:

```
full data set shape for X and y:
(41188, 61) (41188,)

ExtraTreesClassifier(train, test, f1_score):
(1.0, 0.9009420219481402, 0.9009420219481401)

RandomForestClassifier(train, test, f1_score):
(0.9999676281117478, 0.9121103233951636, 0.9121103233951636)

sample data set shape for X and y:
(4000, 61) (4000,)

KNeighborsClassifier(train, test, f1_score):
(0.9323333333333333, 0.916, 0.916)

SVC(train, test, f1_score):
(0.9376666666666666, 0.92, 0.92)

true => no: 902, yes: 98

knn classification report:
predict 'no': 869 (51 misclassifed)
predict 'yes': 47 (33 misclassifed)

svm classification report:
predict 'no': 883 (61 misclassifed)
predict 'yes': 37 (19 misclassifed)
```

Listing 2-6 also displays Figures 2-3 and 2-4. Figure 2-3 displays the confusion matrix for KNeighborsClassifier and Figure 2-4 displays the confusion matrix for svm.SVC.

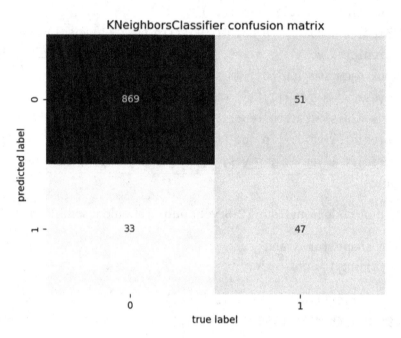

Figure 2-3. *KNeighborsClassifier confusion matrix*

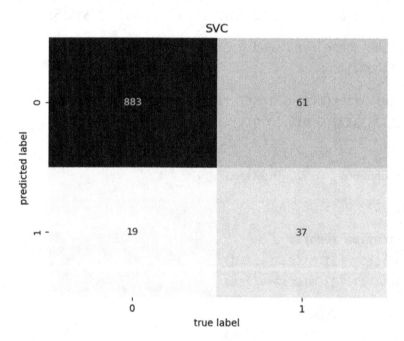

Figure 2-4. *svm.SVC confusion matrix*

The code begins by importing requisite packages. Function get_scores returns train and test accuracy scores. Function prep_data converts NumPy matrices to lists of vectors for easier manipulation of data elements for sampling. Function create_sample builds a random sample and returns it as X and y NumPy matrices.

Scikit-Learn algorithms can only train data represented as NumPy. The main block loads X and y from NumPy files created in the previous example. X and y are split into train-test subsets. The code then trains data with ExtraTreesClassifier and RandomForestClassifier. A sample of 4000 is drawn so that we can efficiently train with KNeighborsClassifier and svm.SVC. These two algorithms are excellent classifiers, but are computationally expensive with large data sets.

Confusion matrices for KNeighborsClassifier and svm.SVC are then displayed because they fit the data better. That is, accuracy was better and there was less overfitting with these models. The code concludes by calculating the balance of target values for the data and misclassifications by KNeighborsClassifier and svm.

Of note is that KNeighborsClassifier and svm.SVC performed better than the other algorithms based on a sample *less than 10%* of the original data. This is actually very impressive!

The UCI Machine Learning Repository includes a randomly selected sample from the bank data with 10% of the examples. For completeness, the next example shown in Listing 2-7 tests accuracy on this sample.

Listing 2-7. Classifying UCI Irvine sample bank data

```
import pandas as pd, numpy as np
from sklearn.model_selection import train_test_split
from sklearn.preprocessing import StandardScaler
from sklearn.ensemble import RandomForestClassifier,\
    ExtraTreesClassifier
from sklearn.svm import SVC
from sklearn.neighbors import KNeighborsClassifier
from sklearn.metrics import f1_score

def get_scores(model, xtrain, ytrain, xtest, ytest, scoring):
    ypred = model.predict(xtest)
    train = model.score(xtrain, ytrain)
    test = model.score(xtest, y_test)
    f1 = f1_score(ytest, ypred, average=scoring)
    return (train, test, f1)
```

```python
if __name__ == "__main__":
    br = '\n'
    f = 'data/bank_sample.csv'
    data = pd.read_csv(f)
    print ('data shape:', data.shape, br)
    data['education'] =\
                        np.where(data['education'] == 'basic.9y',
                                 'basic', data['education'])
    data['education'] = np.where(data['education'] == 'basic.6y',
                                 'basic', data['education'])
    data['education'] = np.where(data['education'] == 'basic.4y',
                                 'basic', data['education'])
    data['education'] = np.where(data['education'] == 'high.school',
                                 'high_school', data.education)
    data['education'] = np.where(data['education'] == 'professional.course',
                                 'professional', data['education'])
    data['education'] = np.where(data['education'] == 'university.degree',
                                 'university', data['education'])
    data_X = data.loc[:, data.columns != 'y']
    cat_vars = ['job', 'marital', 'education', 'default', 'housing',
                'loan', 'contact', 'month', 'day_of_week', 'poutcome']
    data_new = pd.get_dummies(data_X, columns=cat_vars)
    attributes = list(data_X)
    y = data.y.values
    X = data_new.values
    X_train, X_test, y_train, y_test = train_test_split(X, y, random_state=0)
    rf = RandomForestClassifier(random_state=0, n_estimators=100)
    rf.fit(X_train, y_train)
    rf_name = rf.__class__.__name__
    rf_scores = get_scores(rf, X_train, y_train, X_test, y_test, 'micro')
    print (rf.__class__.__name__ + '(train, test, f1_score):')
    print (rf_scores, br)
    et = ExtraTreesClassifier(random_state=0, n_estimators=100)
    et.fit(X_train, y_train)
```

```
et_name = et.__class__.__name__
et_scores = get_scores(et, X_train, y_train, X_test, y_test, 'micro')
print (et.__class__.__name__ + '(train, test, f1_score):')
print (et_scores, br)
scaler = StandardScaler().fit(X_train)
X_train_std, X_test_std = scaler.transform(X_train),\
                          scaler.transform(X_test)
knn = KNeighborsClassifier().fit(X_train, y_train)
knn_scores = get_scores(knn, X_train, y_train, X_test, y_test, 'micro')
print (knn.__class__.__name__ + '(train, test, f1_score):')
print (knn_scores, br)
svm = SVC(random_state=0, gamma='scale')
svm.fit(X_train_std, y_train)
svm_scores = get_scores(svm, X_train_std, y_train, X_test_std, y_test,
                        'micro')
print (svm.__class__.__name__ + '(train, test, f1_score):')
print (svm_scores)
```

After executing code from Listing 2-7, your output should resemble the following:

```
data shape: (4119, 21)

RandomForestClassifier(train, test, f1_score):
(1.0, 0.9058252427184466, 0.9058252427184466)

ExtraTreesClassifier(train, test, f1_score):
(1.0, 0.8990291262135922, 0.8990291262135922)

KNeighborsClassifier(train, test, f1_score):
(0.9323405632890903, 0.8883495145631068, 0.8883495145631068)

SVC(train, test, f1_score):
(0.9494982194885077, 0.9, 0.9)
```

The code begins by importing requisite packages. Function get_scores returns accuracy scores. The main block loads the sample, engineers the *education* feature, and encodes categorical features to OHE form. We had to feature engineer *education* for this example because we didn't draw the sample from the full data set upon which we had already engineered the feature.

The code continues by loading NumPy data into X and y, splitting it into train-test subsets, and training with RandomForestClassifier. Accuracy is then displayed. The remainder of the code trains with ExtraTreesClassifier, KNeighborsClassifier, and svm. SVC and displays accuracy scores.

The sample we created performed at least as well as the one from the UCI repository. Our sample was even a bit smaller, which means that our sampling technique is more than adequate.

Classifying make_moons

Scikit-Learn make_moons data is used primarily to visualize clustering and classification algorithms. However, it is also a great data set to get a sense of how classification algorithms attempt to separate binary target labels (or binary classification). Deployment of make_moons describes two interleaving circles with associated data points in 2D space.

Through visualization, we can easily see the separation between the two (or binary) labels. If human eyes can easily differentiate such separation in 2D space, classification algorithms should be able to do the same. We can test this with an example.

The first code example shown in Listing 2-8 creates a data set with 1000 elements, places feature data and its associated target into a Pandas DataFrame, and plots the result. Each feature element represents an x and y coordinate for plotting in 2D space. Each target represents the feature's label, which is a binary value of either 0 or 1.

Listing 2-8. Plot make_moons

```
import matplotlib.pyplot as plt, pandas as pd
from sklearn import datasets

if __name__ == "__main__":
    br = '\n'
    X, y = datasets.make_moons(n_samples=1000, shuffle=True, noise=0.2,
                               random_state=0)
    df = pd.DataFrame(dict(x=X[:,0], y=X[:,1], label=y))
    colors = {0:'magenta', 1:'cyan'}
    fig, ax = plt.subplots()
    data = df.groupby('label')
    for key, label in data:
```

```
        label.plot(ax=ax, kind='scatter', x='x', y='y', label=key,
                   color=colors[key])
    plt.show()
```

After executing code from Listing 2-8, your output should resemble the following visualization shown in Figure 2-5:

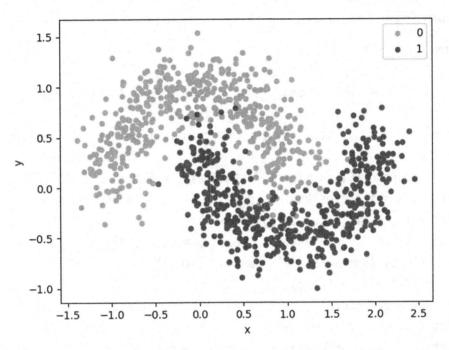

Figure 2-5. *Visualization of randomly generated make_moons data*

The next code example shown in Listing 2-9 creates a make_moons data set of 1000 elements, splits it into train-test subsets, and trains with svm.SVC and KNeighborsClassifier. I intentionally picked these two algorithms because I knew they would do a great job of binary classification since they look at every data point.

Listing 2-9. Classify make_moons

```
from sklearn import datasets
from sklearn.neighbors import KNeighborsClassifier
from sklearn import svm
from sklearn.model_selection import train_test_split
from sklearn.metrics import accuracy_score
```

```
def get_scores(model, Xtrain, Xtest, ytrain, ytest):
    y_ptrain = model.predict(Xtrain)
    y_ptest = model.predict(Xtest)
    acc_train = accuracy_score(ytrain, y_ptrain)
    acc_test = accuracy_score(ytest, y_ptest)
    name = model.__class__.__name__
    return (name, acc_train, acc_test)

if __name__ == "__main__":
    br = '\n'
    X, y = datasets.make_moons(n_samples=1000, shuffle=True, noise=0.2,
                               random_state=0)
    X_train, X_test, y_train, y_test = train_test_split(X, y, random_
                                                        state=0)
    knn = KNeighborsClassifier().fit(X_train, y_train)
    accuracy = get_scores(knn, X_train, X_test, y_train, y_test)
    print ('<<' + str(accuracy[0]) + '>>')
    print ('train:', accuracy[1], 'test:', accuracy[2], br)
    svm = svm.SVC(gamma='scale', random_state=0)
    svm.fit(X_train, y_train)
    accuracy = get_scores(svm, X_train, X_test, y_train, y_test)
    print ('<<' + str(accuracy[0]) + '>>')
    print ('train:', accuracy[1], 'test:', accuracy[2])
```

After executing code from Listing 2-9, your output should resemble the following:

```
<<KNeighborsClassifier>>
train: 0.9666666666666667 test: 0.964

<<SVC>>
train: 0.9653333333333334 test: 0.96
```

The code example begins by importing requisite packages. Function get_scores returns model name and train and test accuracy scores. The main block begins by loading sample data and splitting it into train-test subsets. It continues by training data with KNeighborsClassifier and svm.SVC and reporting accuracy scores. As expected, both algorithms recognized the labels very accurately with essentially no overfitting.

The final code example shown in Listing 2-10 extends our knowledge by splitting data into train, test, and validate subsets. KNeighborsClassifier is used to train and enable reporting.

Listing 2-10. Classify make_moons on train, validate, and test subsets

```
from sklearn.datasets import make_moons
from sklearn.neighbors import KNeighborsClassifier
from sklearn.model_selection import train_test_split
from sklearn.metrics import accuracy_score

def get_scores(model, Xtrain, ytrain, Xtest, ytest, Xvalid, yvalid):
    y_ptrain = model.predict(Xtrain)
    y_ptest = model.predict(Xtest)
    y_pvalid = model.predict(Xvalid)
    acc_train = accuracy_score(ytrain, y_ptrain)
    acc_test = accuracy_score(ytest, y_ptest)
    acc_valid = accuracy_score(yvalid, y_pvalid)
    name = model.__class__.__name__
    return (name, acc_train, acc_test, acc_valid)

if __name__ == "__main__":
    br = '\n'
    X_train, y_train = make_moons(n_samples=1000, shuffle=True, noise=0.2,
                                  random_state=0)
    X_test, y_test = make_moons(n_samples=1000, shuffle=True, noise=0.2,
                                random_state=0)
    X_valid, y_valid = make_moons(n_samples=10000, shuffle=True, noise=0.2,
                                  random_state=0)
    knn = KNeighborsClassifier().fit(X_train, y_train)
    accuracy = get_scores(knn, X_train, y_train, X_test, y_test, X_valid,
                          y_valid)
    print ('train test valid split (technique 1):')
    print ('<<' + str(accuracy[0]) + '>>')
    print ('train:', accuracy[1], 'test:', accuracy[2], 'valid:', accuracy[3])
    print ('sample split:', X_train.shape, X_test.shape, X_valid.shape)
    print ()
```

```
X, y = make_moons(n_samples=1000, shuffle=True, noise=0.2, random_state=0)
X_train, X_test, y_train, y_test = train_test_split(X, y, test_size=0.2,
                                                    random_state=0)
X_train, X_val, y_train, y_val = train_test_split(X_train, y_train,
                                                   test_size=0.25,
                                                   random_state=0)
knn = KNeighborsClassifier().fit(X_train, y_train)
accuracy = get_scores(knn, X_train, y_train, X_test, y_test, X_valid,
                      y_valid)
print ('train test valid split (technique 2):')
print ('<<' + str(accuracy[0]) + '>>')
print ('train:', accuracy[1], 'test:', accuracy[2], 'valid:', accuracy[3])
print ('sample split:', X_train.shape, X_test.shape, X_val.shape)
```

After executing code from Listing 2-10, your output should resemble the following:

```
train test valid split (technique 1):
<<KNeighborsClassifier>>
train: 0.969 test: 0.969 valid: 0.9688
sample split: (1000, 2) (1000, 2) (10000, 2)

train test valid split (technique 2):
<<KNeighborsClassifier>>
train: 0.9616666666666667 test: 0.975 valid: 0.9694
sample split: (600, 2) (200, 2) (200, 2)
```

The code begins importing requisite packages. Function get_scores is expanded to account for validation scores. The main block begins by creating three separate test, train, and validation subsets. With this technique, we create three data sets of the same size. Although this technique produces excellent results, it is much more computationally expensive as data sets become larger and larger. Actually, this technique is three times more expensive because three data sets are created and trained. KNeighborsClassifier is used to train, validate, and test. The second technique is very common because it splits one data set into train, validate, and test. Again, KNeighborsClassifier is used. Results from both techniques are comparable and excellent as expected.

Tip   Test data should only be used once a model is completely trained from training and validation phases so it can provide an unbiased evaluation of a final model fit on training data.

In industry, machine learning engineers experiment with data problems by splitting it into train, test, and validate subsets prior to training. Training data is used to fit (or train) the model. The model sees and learns from training data.

Validation data is used to evaluate a model. Machine learning engineers use validation data to fine-tune the model's hyperparameters. Test data provides an unbiased evaluation of a final model fit based on what was learned from fitting training data and tuning hyperparameters with validation data.

CHAPTER 3

Classification from Complex Training Sets

Classification from complex data is handled exactly as with simple data. Data is loaded into feature set X and target y. X data is composed of a matrix of vectors where each vector represents a data element and y data is composed of a vector of targets. However, complex data is composed of a high number of features (hundreds to thousands). Such a data set is commonly referred to as one with a high-dimensional feature space. Text data is also complex because each document must be converted into vectors of numerical values suitable for machine learning algorithms.

Complex Data Sets

We concentrate on three complex data sets: fetch_20newsgroups, MNIST, and fetch_lfw_people. fetch_20newsgroups is composed of thousands of newsgroup posts (documents). MNIST is composed of thousands of 28 × 28 images where each image is represented by 784 pixels. fetch_lfw_people is composed of 1288 50 × 37 images where each image is represented by 1850 pixels.

Classifying fetch_20newsgroups

Since Scikit-Learn algorithms won't accept raw text, we need to transform it to feature vectors that can be used as input. TfidfVectorizer transforms text (represented as raw documents) into a matrix `54321` of TF-IDF features (feature vectors that can be used as input to an estimator).

TF-IDF (term frequency-inverse document frequency) is a numerical statistic intended to reflect the importance of a word in a document. TF-IDF is one of the most

71

© David Paper 2020
D. Paper, *Hands-on Scikit-Learn for Machine Learning Applications*,
https://doi.org/10.1007/978-1-4842-5373-1_3

popular term-weighting schemes with 83% of text-based recommender system usage in digital libraries.

TF-IDF is a very big topic. We won't go into too much detail because we just want to use it to identify word importance. However, it is important to know that word importance is determined by the TF-IDF weight and importance increases proportionally to the number of times a word appears in a document.

The problem with just looking at word frequency is that some words like "the," "is," and "of" may not be important. So, we can also look at the inverse document frequency, which decreases the weight for commonly used words and increases the weight for words that are not used very much. Fortunately, Scikit-Learn includes the *TfidfVectorizer* package that efficiently combines word and inverse word frequency to extract meaningful information.

Text transformation into feature data differs from images. Image transformation involves flattening matrices into feature vectors of the same length. With text, each document typically differs in size. So, we need a technique like TF-IDF to transform a document into a matrix of TF-IDF features acceptable to Scikit-Learn algorithms. Text transformation is also more complex because word counts impact word importance in a document.

Text classification features are related to word counts or frequencies, so let's use a classifier suited to this purpose. MultinomialNB is a naïve Bayes classifier for multinomial models suitable for classification with discrete features such as word counts for text classification.

The first code example shown in Listing 3-1 classifies fetch_20newsgroups data. You may have to wait a bit the first time you load this data set, so be patient. After the first loading, you won't experience delays.

Listing 3-1. Classify fetch_20newsgroups data

```
from sklearn.datasets import fetch_20newsgroups
from sklearn.feature_extraction.text import TfidfVectorizer
from sklearn.naive_bayes import MultinomialNB
from sklearn.pipeline import make_pipeline
from sklearn.metrics import confusion_matrix, f1_score
import matplotlib.pyplot as plt
import seaborn as sns
```

```python
def predict_category(s, m, t):
    pred = m.predict([s])
    return t[pred[0]]

if __name__ == "__main__":
    br = '\n'
    train = fetch_20newsgroups(subset='train')
    test = fetch_20newsgroups(subset='test')
    print (train.target_names, br)
    categories = ['rec.autos', 'rec.motorcycles', 'sci.space', 'sci.med']
    train = fetch_20newsgroups(subset='train', categories=categories)
    test = fetch_20newsgroups(subset='test', categories=categories)
    print ('data subset:')
    print (train.target.shape, 'shape of train data')
    print (test.target.shape, 'shape of test data', br)
    targets = train.target_names
    mnb_clf = make_pipeline(TfidfVectorizer(), MultinomialNB())
    print ('<<' + mnb_clf.__class__.__name__ + '>>', br)
    mnb_clf.fit(train.data, train.target)
    labels = mnb_clf.predict(test.data)
    f1 = f1_score(test.target, labels, average='micro')
    print ('f1_score', f1, br)
    cm = confusion_matrix(test.target, labels)
    plt.figure('confusion matrix')
    sns.heatmap(cm.T, square=True, annot=True, fmt='d', cmap='gist_ncar_r',
                xticklabels=train.target_names, yticklabels=train.target_
                names, cbar=False)
    print ('sci.med predictions:')
    print (cm.T[2][2], 'correct predictions')
    print (cm.T[2][0], 'misclassified as rec.autos')
    print (cm.T[2][3], 'misclassified as sci.space')
    plt.xlabel('true label')
    plt.ylabel('predicted label')
    plt.tight_layout()
    print ('\n***PREDICTIONS***:')
    y_pred = predict_category('payload on the mars rover', mnb_clf, targets)
```

```
print (y_pred)
y_pred = predict_category('car broke down on the highway', mnb_clf,
                          targets)
print (y_pred)
y_pred = predict_category('dad died of cancer', mnb_clf, targets)
print (y_pred)
```

Go ahead and execute the code from Listing 3-1. Remember that you can find the example from the book's example download. You don't need to type the example by hand. It's easier to access the example download and copy/paste.

Your output from executing Listing 3-1 should resemble the following:

```
['alt.atheism', 'comp.graphics', 'comp.os.ms-windows.misc', 'comp.sys.ibm.
pc.hardware', 'comp.sys.mac.hardware', 'comp.windows.x', 'misc.forsale',
'rec.autos', 'rec.motorcycles', 'rec.sport.baseball', 'rec.sport.hockey',
'sci.crypt', 'sci.electronics', 'sci.med', 'sci.space', 'soc.religion.
christian', 'talk.politics.guns', 'talk.politics.mideast', 'talk.politics.
misc', 'talk.religion.misc']

data subset:
(2379,) shape of train data
(1584,) shape of test data

<<Pipeline>>

f1_score 0.9621212121212122

sci.med predictions:
370 correct predictions
1 misclassified as rec.autos
7 misclassified as sci.space

***PREDICTIONS***:
sci.space
rec.autos
sci.med
```

Listing 3-1 also displays Figure 3-1. Figure 3-1 shows the confusion matrix for the MultinomialNB classification with TfidfVectorizer text transformation.

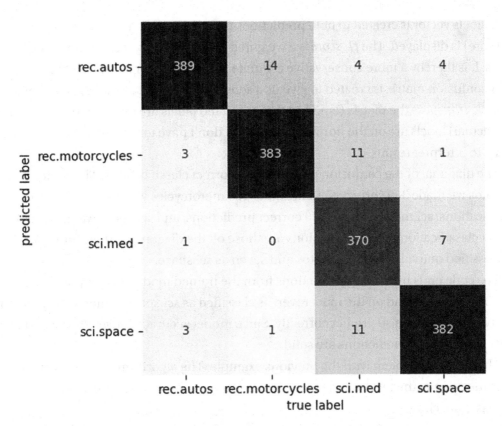

Figure 3-1. *Confusion matrix for the classification experiment*

The code begins by importing fetch_20newsgroups, TfidfVectorizer, MultinomialNB, and a few other familiar packages. Function predict_category is used to predict class label from new data. The main block begins by loading train and test documents from fetch_20newsgroups. Next, it displays the target categories from the data set. The code continues by creating a subset of train and test data with four categories that I chose to use for this experiment. Keep in mind that you can create your *own* subsets.

The subset train and test data sets are used for modeling. Data shapes are then displayed. Next, a pipeline model is created with TfidfVectorizer and MultinomialNB. TfidfVectorizer extracts text and turns it into vectors of numerical values so that MultinomialNB can train it.

Tip   Text data must be converted into numerical form for algorithmic processing.

A labels vector is created to hold predictions from the test data subset. Accuracy score (f1_score) is displayed. The *f1_score* is a weighted average of the precision and recall scores. It is thereby a more conservative estimate. A f1_score over 96% is really good!

A confusion matrix is created to give us a sense of how well our model classified test data. We *transpose* the matrix (cm.T) so that predicted labels are on the vertical axis and true (actual) labels are on the horizontal axis. You don't have to transpose, but it is easier for me to interpret results.

The diagonal of the confusion matrix shows correct classifications. That is, for rec.autos we made 389 correct classifications, rec.motorcycles we made 383 correct classifications, sci.med we made 370 correct predictions, and sci.space we made 383 correct classification. Misclassifications are those off the diagonal. For example, we misclassified one sci.med as rec.autos and seven as sci.space.

The code ends by making predictions from the trained model on completely new data. The text "payload on the mars rover" is classified as sci.space, which is correct! The next two text strings also predict correctly. Since model accuracy is over 96%, we can be pretty sure that our predictions are solid.

There is one problem with the previous example. The algorithm is able to get a sense of the meaning of the text from headers, footers, and quotes. That is, the algorithm we chose is quite clever.

To create a more realistic example, we can remove headers, footers, and quotes from the text documents as shown in the next example in Listing 3-2.

Listing 3-2. Classify fetch_20newsgroups removing identifying information

```
from sklearn.datasets import fetch_20newsgroups
from sklearn.feature_extraction.text import TfidfVectorizer
from sklearn.naive_bayes import MultinomialNB
from sklearn.pipeline import make_pipeline
from sklearn.metrics import confusion_matrix, f1_score
import matplotlib.pyplot as plt
import seaborn as sns

def predict_category(s, m, t):
    pred = m.predict([s])
    return t[pred[0]]
```

```python
if __name__ == "__main__":
    br = '\n'
    train = fetch_20newsgroups(subset='train')
    test = fetch_20newsgroups(subset='test')
    categories = ['rec.autos', 'rec.motorcycles', 'sci.space', 'sci.med']
    train = fetch_20newsgroups(subset='train', categories=categories,
                               remove=('headers', 'footers', 'quotes'))
    test = fetch_20newsgroups(subset='test', categories=categories,
                              remove=('headers', 'footers', 'quotes'))
    targets = train.target_names
    mnb_clf = make_pipeline(TfidfVectorizer(), MultinomialNB())
    print ('<<' + mnb_clf.__class__.__name__ + '>>', br)
    mnb_clf.fit(train.data, train.target)
    labels = mnb_clf.predict(test.data)
    f1 = f1_score(test.target, labels, average='micro')
    print ('f1_score', f1, br)
    cm = confusion_matrix(test.target, labels)
    plt.figure('confusion matrix')
    sns.heatmap(cm.T, square=True, annot=True, fmt='d', cmap='gist_ncar_r',
                xticklabels=train.target_names, yticklabels=train.target_
                names, cbar=False)
    plt.xlabel('true label')
    plt.ylabel('predicted label')
    plt.tight_layout()
    print ('***PREDICTIONS***:')
    y_pred = predict_category('payload on the mars rover', mnb_clf, targets)
    print (y_pred)
    y_pred = predict_category('car broke down on the highway', mnb_clf, targets)
    print (y_pred)
    y_pred = predict_category('dad died of cancer', mnb_clf, targets)
    print (y_pred)
    plt.show()
```

Your output from executing Listing 3-2 should resemble the following:

```
<<Pipeline>>
```

```
f1_score 0.8440656565656567
```

```
***PREDICTIONS***:
sci.space
rec.autos
sci.med
```

Listing 3-2 also displays Figure 3-2. Figure 3-2 shows the confusion matrix once headers, footers, and quotes are removed.

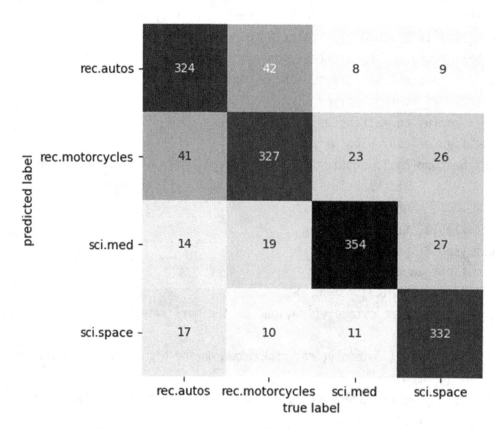

Figure 3-2. *Confusion matrix without headers, footers, and quotes*

The code is very similar to the previous example except we remove headers, footers, and quotes from both train and test subsets. Notice that the f1_score dropped to a bit over 84%, which is quite a big drop! This is a more realistic scenario because text data may not include identifying information.

It appears that predictions from new data are correct, but notice that the text I predict from are pretty clear. That is, it is pretty easy for our trained model to make the correct prediction because the text has words in it that point directly to the right category. For instance, *payload on the mars rover* is definitely *sci.space* because the phrase contains the word *mars*.

The confusion matrix is a great visual in this case because it shows that there are a lot of misclassifications, especially trying to distinguish between rec.autos and rec. motorcycles. Of course, this makes a lot of sense because cars and motorcycles are much more alike than the other two categories in our subset.

Misclassifications are also a great place to search for improvements in algorithms and data. Maybe more data is required to increase accuracy. Maybe the algorithm is misclassifying because some or all of its hyperparameters need adjustment.

Classifying MNIST

MNIST was introduced in Chapter 1 as a large database of handwritten digits commonly used for training and testing in the machine learning community and other industrial image processing applications. As review, MNIST contains 70000 handwritten digit images labeled from 0 to 9 of size 28 × 28. Each target is stored as a digit value. The feature set is a matrix of 70000 28 × 28 images automatically flattened to 784 pixels each.

Training with the Entire MNIST Data Set

The next two code examples train the entire MNIST data set. Since MNIST data consists of high-dimensional feature space, we only train it with select classifiers to reduce training time.

The first code example shown in Listing 3-3 trains MNIST data with RandomForestClassifier and ExtraTreesClassifier, compares accuracy scores, visualizes confusion matrices, and visualizes a misclassification scenario.

Listing 3-3. Classify MNIST data

```python
import numpy as np, humanfriendly as hf
import time
from sklearn.model_selection import train_test_split
from sklearn.ensemble import RandomForestClassifier,\
    ExtraTreesClassifier
from sklearn.metrics import accuracy_score
from sklearn.metrics import confusion_matrix
import matplotlib.pyplot as plt
import seaborn as sns

def get_time(time):
    return hf.format_timespan(time, detailed=True)

def find_misses(test, pred):
    return [i for i, row in enumerate(test) if row != pred[i]]

if __name__ == "__main__":
    br = '\n'
    X_file = 'data/X_mnist'
    y_file = 'data/y_mnist'
    X = np.load('data/X_mnist.npy')
    y = np.load('data/y_mnist.npy')
    X = X.astype(np.float32)
    X_train, X_test, y_train, y_test = train_test_split\
                                    (X, y, random_state=0)
    rf = RandomForestClassifier(random_state=0, n_estimators=100)
    rf_name = rf.__class__.__name__
    print ('<<' + rf_name + '>>')
    start = time.perf_counter()
    rf.fit(X_train, y_train)
    end = time.perf_counter()
    elapsed_ls = end - start
    timer = get_time(elapsed_ls)
    rf_name = rf.__class__.__name__
    y_pred = rf.predict(X_test)
```

```python
accuracy = accuracy_score(y_test, y_pred)
print ('\'test\' accuracy:', accuracy)
print (rf_name + ' timer:', timer, br)
cm = confusion_matrix(y_test, y_pred)
plt.figure(1)
ax = plt.axes()
sns.heatmap(cm.T, annot=True, fmt="d", cmap='gist_ncar_r', ax=ax)
ax.set_title(rf_name + 'confustion matrix')
plt.xlabel('true value')
plt.ylabel('predicted value')
et = ExtraTreesClassifier(random_state=0, n_estimators=100)
et_name = et.__class__.__name__
print ('<<' + et_name + '>>')
start = time.perf_counter()
et.fit(X_train, y_train)
end = time.perf_counter()
elapsed_ls = end - start
timer = get_time(elapsed_ls)
y_pred = et.predict(X_test)
accuracy = accuracy_score(y_test, y_pred)
print ('\'test\' accuracy:', accuracy)
print (et_name + ' timer:', timer, br)
cm = confusion_matrix(y_test, y_pred)
plt.figure(2)
ax = plt.axes()
sns.heatmap(cm.T, annot=True, fmt="d", cmap='gist_ncar_r', ax=ax)
ax.set_title(et_name + 'confustion matrix')
plt.xlabel('true value')
plt.ylabel('predicted value')
indx = find_misses(y_test, y_pred)
print ('pred', 'actual')
misses = [(y_pred[row], y_test[row], i)
          for i, row in enumerate(indx)]
[print (row[0], '   ', row[1]) for i, row in enumerate(misses)
 if i < 5]
print()
```

```
img_act = y_test[indx[0]]
img_pred = y_pred[indx[0]]
print ('actual', img_act)
print ('pred', img_pred)
text = str(img_pred)
test_images = X_test.reshape(-1, 28, 28)
plt.figure(3)
plt.imshow(test_images[indx[0]], cmap='gray', interpolation='gaussian')
plt.text(0, 0.05, text, color='r', bbox=dict(facecolor='white'))
title = str(img_act) + ' misclassified as ' + text
plt.title(title)
plt.show()
```

Your output from executing Listing 3-3 should resemble the following:

```
<<RandomForestClassifier>>
'test' accuracy: 0.9687428571428571
RandomForestClassifier timer: 29 seconds and 620.77 milliseconds

<<ExtraTreesClassifier>>
'test' accuracy: 0.9727428571428571
ExtraTreesClassifier timer: 30 seconds and 462.9 milliseconds

pred actual
3.0     9.0
7.0     3.0
4.0     9.0
2.0     3.0
3.0     9.0

actual 9.0
pred 3.0
```

Listing 3-3 also displays Figures 3-3, 3-4, and 3-5. Figure 3-3 shows the confusion matrix for RandomForestClassifier. Figure 3-4 shows the confusion matrix for ExtraTreesClassifier. Figure 3-5 show the first misclassification.

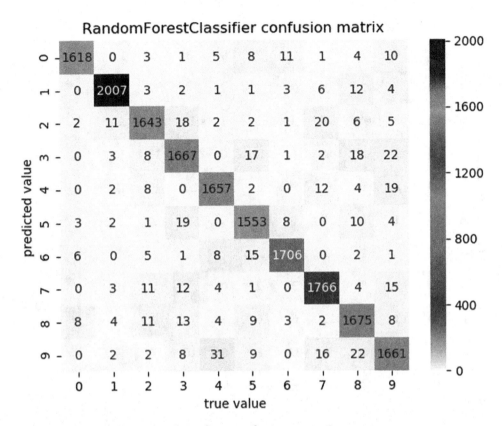

Figure 3-3. *RandomForestClassifier confusion matrix*

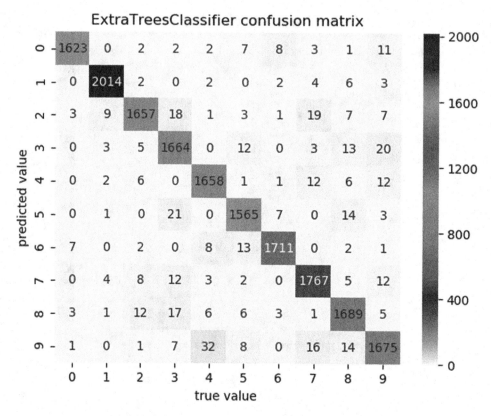

Figure 3-4. *ExtraTreesClassifier confusion matrix*

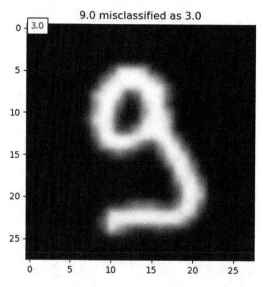

Figure 3-5. *The first misclassification*

The code begins by importing requisite packages. Function get_time returns the time it takes to train. Function find_misses returns a list of misclassifications. The main block loads MNIST from NumPy files into X and y, converts X into float for algorithmic consumption, and splits data into train-test subsets.

The code continues by training data with RandomForestClassifier and ExtraTreesClassifier. For both algorithms, accuracy and training time are displayed, and confusion matrices are created and visualized. Next, the first five misclassifications are displayed. Finally, the first misclassification (pred: 3.0, actual: 9.0) is visualized. Notice that it does take some time for training (approximately 30 seconds for each algorithm). The reason is that MNIST data is composed of high-dimensional feature space.

Notice that it is easy for humans to see that the digit is 9, but not so easy for a machine because the prediction was digit 3. This visualization as well as the confusion matrices is really important because it can help data scientists improve prediction performance by adjusting the data, experimenting with different algorithms, or improving the algorithms.

The next code example shown in Listing 3-4 splits data into train-test subsets manually for increased flexibility. That is, we can granularly adjust train-test subset sizes.

Listing 3-4. Classify MNIST with manual train-test shuffle

```
import numpy as np, humanfriendly as hf
import time
from sklearn.ensemble import ExtraTreesClassifier
from sklearn.metrics import accuracy_score
from sklearn.metrics import classification_report

if __name__ == "__main__":
    br = '\n'
    X_file = 'data/X_mnist'
    y_file = 'data/y_mnist'
    X = np.load('data/X_mnist.npy')
    y = np.load('data/y_mnist.npy')
    X = X.astype(np.float32)
    X_train, X_test, y_train, y_test = X[:60000], X[60000:],\
                                       y[:60000], y[60000:]
    shuffle_index = np.random.permutation(60000)
```

```
X_train, y_train = X_train[shuffle_index],\
                   y_train[shuffle_index]
et = ExtraTreesClassifier(random_state=0, n_estimators=100)
start = time.perf_counter()
et.fit(X_train, y_train)
end = time.perf_counter()
elapsed_ls = end - start
print (hf.format_timespan(elapsed_ls, detailed=True))
et_name = et.__class__.__name__
y_pred = et.predict(X_test)
accuracy = accuracy_score(y_test, y_pred)
print (et_name + ' \'test\':', end=' ')
print ('accuracy:', accuracy, br)
rpt = classification_report(y_test, y_pred)
print (rpt)
```

Your output from executing Listing 3-4 should resemble the following:

```
36 seconds and 533.76 milliseconds
ExtraTreesClassifier 'test': accuracy: 0.9706
```

	precision	recall	f1-score	support
0.0	0.97	0.99	0.98	980
1.0	0.99	0.99	0.99	1135
2.0	0.97	0.97	0.97	1032
3.0	0.97	0.96	0.96	1010
4.0	0.97	0.97	0.97	982
5.0	0.97	0.97	0.97	892
6.0	0.98	0.98	0.98	958
7.0	0.97	0.97	0.97	1028
8.0	0.97	0.96	0.96	974
9.0	0.95	0.95	0.95	1009
micro avg	0.97	0.97	0.97	10000
macro avg	0.97	0.97	0.97	10000
weighted avg	0.97	0.97	0.97	10000

The code begins by importing classification_report and other requisite packages. The classification report displays precision, recall, F1, and support scores for the model.

Precision is the ability of a classifier not to label an instance positive that is actually negative. *Recall* is the ability of a classifier to find all positive instances. F_1_score is a weighted harmonic mean of precision and recall where the best score is 1.0 and the worst is 0.0. F_1_scores are generally lower than accuracy measures as they embed precision and recall into their computation. *Support* is the number of actual occurrences of the class in the specified data set.

The main block loads MNIST into X and y and manually shuffles data into train-test subsets. In this case, we used more data for training. Specifically, 60000 for training and 10000 for testing. We have more training data, which can be a good experiment if we have enough data to work with. Next, we trained with ExtraTreesClassifier since it performed better than RandomForestClassifier on MNIST in the previous example. Finally, accuracy score and a classification report are presented.

Tip   Manually shuffle data into train-test subsets if you desire more flexibility.

Training MNIST Sample Data

The first code example shown in Listing 3-5 creates a random sample of 4000 data elements from MNIST to enable efficient training with svm.SVC and KNeighborsClassifier. Both of these algorithms are excellent classifiers, but they are known to be computationally expensive with high-dimensional feature space data sets.

Listing 3-5. Classify MNIST with sample data

```
import numpy as np, random, humanfriendly as hf
import time
from sklearn.model_selection import train_test_split
from sklearn.preprocessing import StandardScaler
from sklearn import svm
from sklearn.neighbors import KNeighborsClassifier
import matplotlib.pyplot as plt
```

```python
def prep_data(data, target):
    d = [data[i] for i, _ in enumerate(data)]
    t = [target[i] for i, _ in enumerate(target)]
    return list(zip(d, t))

def create_sample(d, n, replace='yes'):
    if replace == 'yes': s = random.sample(d, n)
    else: s = [random.choice(d)
                    for i, _ in enumerate(d) if i < n]
    Xs = [row[0] for i, row in enumerate(s)]
    ys = [row[1] for i, row in enumerate(s)]
    return np.array(Xs), np.array(ys)

def see_time(note):
    end = time.perf_counter()
    elapsed = end - start
    print (note, hf.format_timespan(elapsed, detailed=True))

if __name__ == "__main__":
    br = '\n'
    X_file = 'data/X_mnist'
    y_file = 'data/y_mnist'
    X = np.load('data/X_mnist.npy')
    y = np.load('data/y_mnist.npy')
    X = X.astype(np.float32)
    sample_size = 4000
    data = prep_data(X, y)
    Xs, ys = create_sample(data, sample_size, replace='no')
    X_train, X_test, y_train, y_test = train_test_split(
        Xs, ys, test_size=0.10, random_state=0)
    scaler = StandardScaler().fit(X_train)
    X_train_std, X_test_std = scaler.transform(X_train),\
                                    scaler.transform(X_test)
    svm = svm.SVC(random_state=0, gamma='scale')
    svm_name = svm.__class__.__name__
    print ('<<', svm_name, '>>')
    start = time.perf_counter()
```

```
svm.fit(X_train_std, y_train)
see_time('train:')
start = time.perf_counter()
y_pred = svm.predict(X_test_std)
see_time('predict:')
start = time.perf_counter()
train_score = svm.score(X_train_std, y_train)
test_score = svm.score(X_test_std, y_test)
see_time('score:')
print ('train score:', train_score, 'test score', test_score, br)
knn = KNeighborsClassifier()
knn_name = knn.__class__.__name__
print ('<<', knn_name, '>>')
start = time.perf_counter()
knn.fit(X_train, y_train)
see_time('train:')
start = time.perf_counter()
y_pred = knn.predict(X_test)
see_time('predict:')
start = time.perf_counter()
train_score = knn.score(X_train, y_train)
test_score = knn.score(X_test, y_test)
see_time('score:')
print ('train score:', train_score, 'test score:', test_score)
```

Your output from executing Listing 3-5 should resemble the following:

```
train: 6 seconds and 538.51 milliseconds
predict: 780.46 milliseconds
score: 7 seconds and 755.28 milliseconds
train score: 0.9802777777777778 test score 0.9075

<< KNeighborsClassifier >>
train: 116.53 milliseconds
predict: 1 second and 605.23 milliseconds
score: 15 seconds and 924.84 milliseconds
train score: 0.9519444444444445 test score: 0.91
```

The code begins by importing requisite packages. Function prep_data transforms data into a list of data elements for easier sampling. Function create_sample accepts prepared data and creates a random sample without replacement. Function see_time returns elapsed time.

The main block loads data into X and y. Next, a sample of 4000 data elements is created and split into train-test subsets. The code continues by training the sample with svm.SVC and KNeighborsClassifier.

We needed to draw a random sample with these algorithms because they are computationally expensive when training large data sets, especially those with a high-dimensional feature space.

For each algorithm, train, predict, and score times are reported. To this point, we have only captured train time. But, it is interesting to see how much time each phase of the training process occupies. For our sample, KNeighborsClassifier reported respectable results with less overfitting than svm.SVC.

Tip   You can easily experiment with sample size by adjusting the sample_size variable.

The next MNIST example shown in Listing 3-6 leverages PCA to allow an increased sample size of 7000 without too much added computational expense.

Listing 3-6. Classify MNIST with sample data and PCA

```
import numpy as np, random, humanfriendly as hf
import time
from sklearn.decomposition import PCA
from sklearn.model_selection import train_test_split
from sklearn.preprocessing import StandardScaler
from sklearn import svm
from sklearn.neighbors import KNeighborsClassifier
from sklearn.metrics import f1_score
from sklearn.metrics import confusion_matrix
import matplotlib.pyplot as plt, seaborn as sns
```

```python
def prep_data(data, target):
    d = [data[i] for i, _ in enumerate(data)]
    t = [target[i] for i, _ in enumerate(target)]
    return list(zip(d, t))

def create_sample(d, n, replace='yes'):
    if replace == 'yes': s = random.sample(d, n)
    else: s = [random.choice(d)
                for i, _ in enumerate(d) if i < n]
    Xs = [row[0] for i, row in enumerate(s)]
    ys = [row[1] for i, row in enumerate(s)]
    return np.array(Xs), np.array(ys)

def see_time(note):
    end = time.perf_counter()
    elapsed = end - start
    print (note, hf.format_timespan(elapsed, detailed=True))

def get_scores(model, xtrain, ytrain, xtest, ytest):
    ypred = model.predict(xtest)
    train = model.score(xtrain, ytrain)
    test = model.score(xtest, y_test)
    f1 = f1_score(ytest, ypred, average='macro')
    return (ypred, train, test, f1)

if __name__ == "__main__":
    br = '\n'
    X_file = 'data/X_mnist'
    y_file = 'data/y_mnist'
    X = np.load('data/X_mnist.npy')
    y = np.load('data/y_mnist.npy')
    X = X.astype(np.float32)
    data = prep_data(X, y)
    sample_size = 7000
    Xs, ys = create_sample(data, sample_size, replace='no')
    pca = PCA(n_components=0.95, random_state=0)
    Xs_reduced = pca.fit_transform(Xs)
    print ('sample feature shape:', Xs.shape)
```

```
components = pca.n_components_
print ('feature components with PCA:', components, br)
X_train, X_test, y_train, y_test = train_test_split(
    Xs_reduced, ys, test_size=0.10, random_state=0)
scaler = StandardScaler().fit(X_train)
X_train_std, X_test_std = scaler.transform(X_train),\
                          scaler.transform(X_test)
start = time.perf_counter()
svm = svm.SVC(random_state=0).fit(X_train_std, y_train)
svm_name = svm.__class__.__name__
svm_scores = get_scores(svm, X_train_std, y_train, X_test_std, y_test)
cm_svm = confusion_matrix(y_test, svm_scores[0])
see_time(svm_name + ' total training time:')
print (svm_name + ':', svm_scores[1], svm_scores[2], svm_scores[3], br)
start = time.perf_counter()
knn = KNeighborsClassifier().fit(X_train, y_train)
knn_name = knn.__class__.__name__
knn_scores = get_scores(knn, X_train, y_train, X_test, y_test)
cm_knn = confusion_matrix(y_test, knn_scores[0])
see_time(knn_name + ' total training time:')
print (knn_name + ':', knn_scores[1], knn_scores[2], knn_scores[3])
plt.figure(svm_name)
ax = plt.axes()
sns.heatmap(cm_svm.T, annot=True, fmt="d", cmap='gist_ncar_r', ax=ax)
ax.set_title(str(svm_name) + ' confustion matrix')
plt.xlabel('true value')
plt.ylabel('predicted value')
plt.figure(knn_name)
ax = plt.axes()
sns.heatmap(cm_knn.T, annot=True, fmt="d", cmap='gist_ncar_r', ax=ax)
ax.set_title(str(knn_name) + ' confustion matrix')
plt.xlabel('true value')
plt.ylabel('predicted value')
plt.show()
```

Your output from executing Listing 3-6 should resemble the following:

```
sample feature shape: (7000, 784)
feature components with PCA: 150

SVC total training time: 14 seconds and 290.91 milliseconds
SVC: 0.9955555555555555 0.9428571428571428 0.9425480948692136

KNeighborsClassifier total training time: 10 seconds and 313.37
milliseconds
KNeighborsClassifier: 0.9601587301587302 0.9371428571428572
0.9358573966927535
```

Listing 3-6 also displays Figures 3-6 and 3-7. Figure 3-6 shows the confusion matrix for svm.SVC. Figure 3-7 shows the confusion matrix for KNeighborsClassifier.

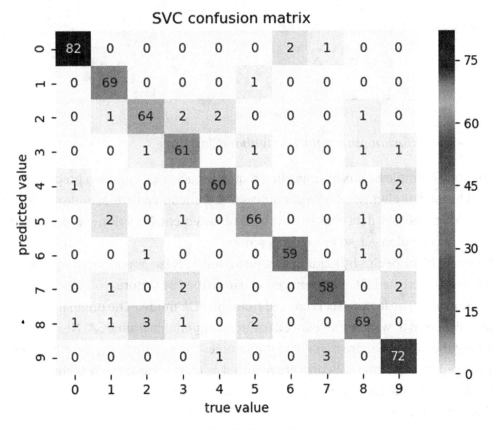

Figure 3-6. *Confusion matrix for svm.SVC*

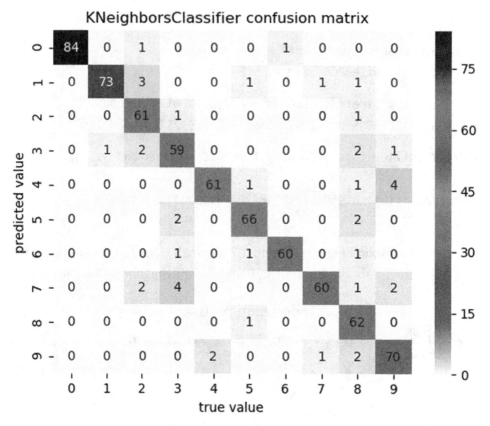

Figure 3-7. *Confusion matrix for KNeighborsClassifier*

The code example begins by importing requisite packages. Function prep_data creates a list of data elements for easier processing. Function create_sample creates a random sample of 7000 data elements without replacement. Function see_time returns elapsed time. Function get_scores returns scores.

The main block begins by loading data into X and y. Next, a random sample of 7000 data elements is created. PCA is leveraged to *reduce* the 784 features to 150 with 5% information loss. The feature set is created from the PCA model. The original sample shape is displayed as well as the reduced feature components from PCA. Next, svm.SVC and KNeighborsClassifier are trained on the sample. The total training time and scoring are displayed for each model. Scores are reported as train accuracy, test accuracy, and test f1_score, respectively. The code concludes by creating and displaying confusion matrices for each model.

Notice that using the larger sample of 7000 resulted in a better fit for both models. That is, we have less overfitting.

Sampling is very common practice in industry as data sets become larger and larger. Sampling can drastically reduce computational expense while providing a good idea of the predictive capability of even the most computationally expensive algorithms. In addition, dimensionality reduction in conjunction with sampling can reduce computational expense even more!

Tip   Dimensionality reduction with sampling can drastically reduce computational expense.

Classifying fetch_lfw_people

The fetch_lfw_people consists of preprocessed images from Labeled Faces in the Wild (LFW), which is a database designed for studying unconstrained face recognition. LFW contains more than 13,000 images of faces collected from the Web. Each image is labeled with the name of the person pictured. To learn more about LFW, follow this link: *http://vis-www.cs.umass.edu/lfw/*. For our experiment, we only consider folks with a minimum of 70 pictures in the data set. Images are resized to a 0.4 aspect ratio.

The first code example shown in Listing 3-7 classifies fetch_lfw_people with svm. SVC, which is one of the most useful algorithms for face recognition.

Listing 3-7. Classify fetch_lfw_people data

```
import numpy as np
from sklearn.decomposition import PCA
from sklearn.model_selection import train_test_split
from sklearn.svm import SVC
from sklearn.metrics import classification_report
import matplotlib.pyplot as plt

if __name__ == "__main__":
    br = '\n'
    X = np.load('data/X_faces.npy')
    y = np.load('data/y_faces.npy')
    images = np.load('data/faces_images.npy')
    targets = np.load('data/faces_targets.npy')
    _, h, w = images.shape
```

```
n_images = X.shape[0]
n_features = X.shape[1]
n_classes = len(targets)
print ('features:', n_features)
print ('images:', n_images)
print ('classes:', n_classes, br)
print ('target names:')
print (targets, br)
X_train, X_test, y_train, y_test = train_test_split(X, y, random_state=0)
pca = PCA(n_components=0.95, whiten=True, random_state=0)
pca.fit(X_train)
components = pca.n_components_
eigenfaces = pca.components_.reshape((components, h, w))
X_train_pca = pca.transform(X_train)
pca_name = pca.__class__.__name__
print ('<<' + pca_name + '>>')
print ('features (after PCA):', components)
print ('eigenface shape:', eigenfaces.shape, br)
print (pca, br)
svm = SVC(kernel='rbf', class_weight='balanced', gamma='scale',
          random_state=0)
svm_name = svm.__class__.__name__
svm.fit(X_train_pca, y_train)
X_test_pca = pca.transform(X_test)
y_pred = svm.predict(X_test_pca)
cr = classification_report(y_test, y_pred)
print ('classification report <<' + svm_name+ '>>')
print (cr)
ls = [np.array(eigenfaces[i].reshape(h, w))
      for i, row in enumerate(range(9))]
fig, ax = plt.subplots(3, 3, figsize=(5, 6))
cnt = 0
for row in [0, 1, 2]:
    for col in [0, 1, 2]:
        ax[row, col].imshow(ls[cnt], cmap='bone', aspect='auto')
```

```
        ax[row, col].set_axis_off()
        cnt += 1
    plt.tight_layout()
    plt.show()
```

Your output from executing Listing 3-7 should resemble the following:

```
features: 1850
images: 1288
classes: 7

target names:
['Ariel Sharon' 'Colin Powell' 'Donald Rumsfeld' 'George W Bush'
 'Gerhard Schroeder' 'Hugo Chavez' 'Tony Blair']

<<PCA>>
features (after PCA): 135
eigenface shape: (135, 50, 37)

PCA(copy=True, iterated_power='auto', n_components=0.95,
  random_state=0, svd_solver='auto', tol=0.0, whiten=True)

classification report <<SVC>>
              precision    recall  f1-score   support

           0       1.00      0.61      0.76        28
           1       0.63      0.94      0.76        63
           2       0.90      0.75      0.82        24
           3       0.88      0.86      0.87       132
           4       0.75      0.75      0.75        20
           5       1.00      0.59      0.74        22
           6       0.90      0.82      0.86        33

   micro avg       0.82      0.82      0.82       322
   macro avg       0.87      0.76      0.79       322
weighted avg       0.85      0.82      0.82       322
```

Listing 3-7 also displays Figure 3-8. Figure 3-8 shows the eigenfaces created by PCA.

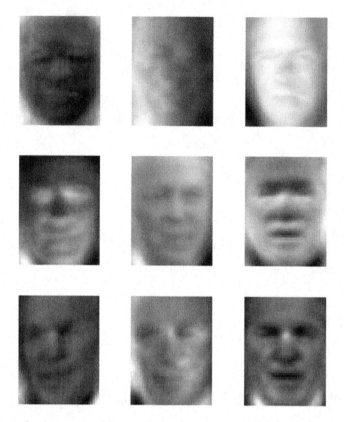

Figure 3-8. *Eigenfaces created by PCA*

The code example begins by loading requisite packages. The main block loads image data into feature set X, target set y, image matrices into variable *images*, and target names into variable *targets*. The code continues by splitting data into train-test subsets. Next, a PCA model is created with whitening set to *True*. Whitening is enabled to mitigate redundancy in the input data (or X_train).

Since each image consists of 1850 pixels, our feature set has 1850 dimensions. So, PCA allows us to reduce dimensions to 135. With fewer dimensions (or features), computational expense is reduced. Fewer features also reduce the complexity of the model, which can mitigate overfitting.

PCA attempts to represent training data variance with as few dimensions as possible by keeping the most important features. When PCA is used with images, the remaining features are commonly called eigenfaces. *Eigenfaces* represent the principal set of images that are projected onto each data example from the train set to obtain independent features. That is, eigenfaces are used by the algorithm to learn from the data.

PCA trains on X_train data with 5% information loss. Next, PCA components and eigenfaces (or best remaining features) are determined. The code continues by transforming X_train with PCA into X_train_pca with 135 features (rather than 1850). We then train X_train_pca with svm and create the prediction set y_pred so we can create a classification report. The code ends by creating images from the first nine data elements from eigenfaces. For further reading, visit:*http://efavdb.com/machine-learning-for-facial-recognition-3/.*

Tip   PCA is not only a good model for unsupervised learning experiments, it enables dimensionality reduction on train sets that result in faster processing and less overfitting for supervised learning experiments.

The next code example shown in Listing 3-8 trains data exactly as in the previous example, but this time we visualize the first correct classification and the first misclassification. The code continues by visualizing four random predictions. The code may seem very complex, but much of the effort is tied to creating nice visuals with Matplotlib. If you haven't already figured it out, Matplotlib is *not* very user friendly.

Listing 3-8. Classify fetch_lfw_people data and visualize

```
import numpy as np
from random import randint
from sklearn.decomposition import PCA
from sklearn.model_selection import train_test_split
from sklearn.svm import SVC
import matplotlib.pyplot as plt

def find_misses(test, pred):
    return [i for i, row in enumerate(test) if row != pred[i]]

def find_hit(n, ls):
    return True if n in ls else False

def build_fig(indx, pos, color, one, two):
    X_i = np.array(X_test[indx]).reshape(50, 37)
    t = targets[y_test[indx]]
    p = targets[y_pred[indx]]
```

```python
    ax = fig.add_subplot(pos)
    image = ax.imshow(X_i,  cmap='bone')
    ax.set_axis_off()
    ax.set_title(t)
    ax.text(one, two, p, color=color, bbox=dict(facecolor='white'))

def chk_acc(rnds):
    logic = [1 if y_test[row] == y_pred[row] else 0 for row in rnds]
    colors = ['g' if row == 1 else 'r' for row in logic]
    return colors

if __name__ == "__main__":
    br = '\n'
    X = np.load('data/X_faces.npy')
    y = np.load('data/y_faces.npy')
    images = np.load('data/faces_images.npy')
    targets = np.load('data/faces_targets.npy')
    X_train, X_test, y_train, y_test = train_test_split(X, y, random_state=0)
    pca = PCA(n_components=0.95, whiten=True, random_state=0)
    pca.fit(X_train)
    X_train_pca = pca.transform(X_train)
    pca_name = pca.__class__.__name__
    svm = SVC(kernel='rbf', class_weight='balanced', gamma='scale',
              random_state=0)
    svm_name = svm.__class__.__name__
    svm.fit(X_train_pca, y_train)
    X_test_pca = pca.transform(X_test)
    y_pred = svm.predict(X_test_pca)
    misses = find_misses(y_test, y_pred)
    miss = misses[0]
    hit = 1
    X_hit = np.array(X_test[hit]).reshape(50, 37)
    y_test_hit = targets[y_test[hit]]
    y_pred_hit = targets[y_pred[hit]]
    X_miss = np.array(X_test[miss]).reshape(50, 37)
    y_test_miss = targets[y_test[miss]]
    y_pred_miss = targets[y_pred[miss]]
```

```
fig = plt.figure('1st Hit and Miss')
fig.suptitle('Visualize 1st Hit and Miss', fontsize=18, fontweight='bold')
build_fig(hit, 121, 'g', 0.4, 1.9)
build_fig(miss, 122, 'r', 0.4, 1.9)
rnd_ints = [randint(0, y_test.shape[0]-1) for row in range(4)]
colors = chk_acc(rnd_ints)
fig = plt.figure('Four Random Predictions')
build_fig(rnd_ints[0], 221, colors[0], .9, 4.45)
build_fig(rnd_ints[1], 222, colors[1], .9, 4.45)
build_fig(rnd_ints[2], 223, colors[2], .9, 4.45)
build_fig(rnd_ints[3], 224, colors[3], .9, 4.45)
plt.tight_layout()
plt.show()
```

Listing 3-8 displays Figures 3-9 and 3-10. Figure 3-9 is a visualization of the first hit (or correct classification) and first miss (misclassification) from the training experiment. Figure 3-10 is a visualization of four random predictions.

Visualize 1st Hit and Miss

Figure 3-9. *First hit and miss from the training experiment*

Hugo Chavez

Colin Powell

Donald Rumsfeld

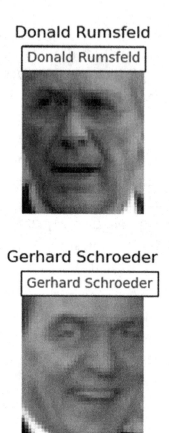

Donald Rumsfeld

George W Bush

George W Bush

Gerhard Schroeder

Gerhard Schroeder

Figure 3-10. *Four random predictions*

The code begins by importing requisite packages. Function find_misses returns the index of misclassifications from the test set. Function find_hit assists in finding whether an index provided was classified correctly.

Function find_hit is not enacted in the code provided, but you can test it yourself by feeding an index and the list of misses into the function. If the function returns *True*, the prediction was correct; otherwise the prediction was misclassified. I tested the function with index 1 and the function returned True.

Function build_figure enables us build the visualizations. Although the code appears complex, it is really just meticulous. That is, it just takes time to position the text properly. Function chk_acc returns red for a misclassification and green for a correct one.

The main block loads data and trains exactly as in the previous example. The remaining code creates the visualizations. The first visual displays the first correct classification and the first misclassification from the test set. So, the Colin Powell image was correctly classified at index 1 (2nd data element in the test set). The George W. Bush image was misclassified as Colin Powell, and it happens to be at index 0 (first data element in the test set).

The second visual was created by generating four random numbers and using them as indices for visualization from the test set. Our visualization shows three of four correct classifications. The only misclassification was Hugo Chavez misclassified as Colin Powell.

Keep in mind that each time you run the code, you will see different images because we randomly generate indices. In addition, you are more than likely to see four correct classifications because our accuracy is 85%.

The final code example shown in Listing 3-9 is included for completeness. I want to show you how to use LDA dimensionality reduction rather than PCA. In this case, LDA performed poorly, but it may perform better given a different data set.

Listing 3-9. Dimensionality reduction with LDA

```
import numpy as np
from sklearn.decomposition import PCA
from sklearn.discriminant_analysis import\
    LinearDiscriminantAnalysis
from sklearn.model_selection import train_test_split
from sklearn.svm import SVC
from sklearn.metrics import classification_report
import warnings

if __name__ == "__main__":
    br = '\n'
    warnings.filterwarnings('ignore')
    X = np.load('data/X_faces.npy')
    y = np.load('data/y_faces.npy')
    X_train, X_test, y_train, y_test = train_test_split(X, y, random_state=0)
    pca = PCA(n_components=0.95, whiten=True, random_state=0)
    pca.fit(X_train)
    components = pca.n_components_
    lda = LinearDiscriminantAnalysis(n_components=components)
    lda.fit(X_train, y_train)
    X_train_lda = lda.transform(X_train)
```

```
svm = SVC(kernel='rbf', class_weight='balanced', gamma='scale',
        random_state=0)
svm_name = svm.__class__.__name__
svm.fit(X_train_lda, y_train)
X_test_lda = lda.transform(X_test)
y_pred = svm.predict(X_test_lda)
cr = classification_report(y_test, y_pred)
print ('classification report <<' + svm_name+ '>>')
print (cr)
```

Your output from executing Listing 3-9 should resemble the following:

```
classification report <<SVC>>
```

	precision	recall	f1-score	support
0	1.00	0.21	0.35	28
1	0.84	0.49	0.62	63
2	0.69	0.38	0.49	24
3	0.56	0.96	0.71	132
4	0.50	0.15	0.23	20
5	0.73	0.36	0.48	22
6	0.67	0.42	0.52	33
micro avg	0.61	0.61	0.61	322
macro avg	0.71	0.43	0.49	322
weighted avg	0.68	0.61	0.58	322

The code is short because the performance is so much lower than PCA (68% vs. 85% accuracy). Why should we create eigenfaces, predictions, and visualizations if PCA is the better model? Notice that we use PCA to determine the best number of components with 5% information loss, which is then used by LDA for prediction with svm.SVC.

CHAPTER 4

Predictive Modeling Through Regression

While classification is the problem of predicting a discrete class label for an example, regression predictive modeling (or just *regression*) is the problem of learning the strength of association between independent variables (or features) and continuous dependent variables (or outcomes). A *continuous* output variable is a real value such as an integer or floating point value often quantified as amounts and sizes.

Simply, regression attempts to learn how strong the relationship is between features and outcomes. Formally, regression approximates a mapping function (f) from input variables (X) to a continuous output variable (y). An algorithm capable of learning a regression predictive model is called a regression algorithm. Since regression predicts a quantity, *the performance must be measured as error* in those predictions.

Performance of regression can be gauged in many ways, but the most common is to calculate root mean squared error (RMSE). A benefit of *RMSE* is that units of the error score are the same as the predicted value. While regression predictions can be evaluated using RMSE, classification predictions cannot.

Regression Data Sets

We concentrate on four data sets: tips, boston, and wine (red and white). tips data is composed of food server tips in restaurants and related factors including tip, price of meal, and time of day. boston data is composed of housing prices from various Boston locations. wine data is composed of two data sets (red and white) that consist of variants of Portuguese Vinho Verde wine.

© David Paper 2020
D. Paper, *Hands-on Scikit-Learn for Machine Learning Applications*,
https://doi.org/10.1007/978-1-4842-5373-1_4

Regressing tips

The first code example shown in Listing 4-1 loads tips data from a CSV file, conducts feature engineering by converting categorical features to dummy features, adds new data to the existing data set, imputes new data, displays feature importance, trains data with the LinearRegression algorithm, and predicts. In this case, we learn from multiple linear regression because we are training data on multiple features and one continuous dependent target.

Listing 4-1. Predicting from tips with get_dummies encoding

```
import numpy as np, pandas as pd
from sklearn.ensemble import RandomForestRegressor
from sklearn.impute import SimpleImputer
from sklearn.linear_model import LinearRegression
from sklearn.model_selection import train_test_split
from sklearn.metrics import mean_squared_error

if __name__ == "__main__":
    br = '\n'
    tips = pd.read_csv('data/tips.csv')
    print ('original data shape:', tips.shape, br)
    target = tips['tip']
    data = tips.drop(['tip'], axis=1)
    data = pd.get_dummies(data, columns=['sex', 'smoker','day', 'time'])
    d = {'sex_Male':'male', 'sex_Female':'female',
         'smoker_Yes':'smoker', 'smoker_No':'non-smoker',
         'day_Thur':'th', 'day_Fri':'fri', 'day_Sat':'sat',
         'day_Sun':'sun', 'time_Lunch':'lunch',
         'time_Dinner':'dinner'}
    data = data.rename(index=str, columns=d)
    X = data.values
    y = target.values
    print ('X and y shapes (post conversion):')
    print (X.shape, y.shape, br)
```

```
X_vector = np.array([30.00, 'NaN', 1, 0, 1, 0, 0, 0, 0, 1, 1, 0])
y_vector = np.array([4.5])
X = np.vstack([X, X_vector])
y = np.append(y, y_vector)
print ('new X and y data point:')
print (X[244], y[244], br)
X_vectors = np.array([[24.99, 'NaN',0, 1, 0, 1, 1, 0, 0, 0, 0, 1],
                      [19.99, 'NaN',1, 0, 1, 0, 0, 0, 0, 1, 1, 0]])
y_vectors = np.array([[3.5], [2.0]])
X = np.vstack([X, X_vectors])
y = np.append(y, y_vectors)
print ('new X and y data points:')
print (X[245], y[245])
print (X[246], y[246], br)
imputer = SimpleImputer()
imputer.fit(X)
X = imputer.transform(X)
print ('new data shape:', X.shape, br)
print ('new records post imputation (features and targets):')
print (X[244], y[244])
print (X[245], y[245])
print (X[246], y[246], br)
rfr = RandomForestRegressor(random_state=0, n_estimators=100)
rfr.fit(X, y)
print ('feature importance (first 6 features):')
feature_importances = rfr.feature_importances_
features = list(data.columns.values)
importance = sorted(zip(feature_importances, features), reverse=True)
[print (row) for i, row in enumerate(importance) if i < 6]
print ()
X_train, X_test, y_train, y_test = train_test_split(
    X, y, random_state=0)
```

```
model = LinearRegression()
model_name = model.__class__.__name__
print ('<<' + model_name + '>>', br)
model.fit(X_train, y_train)
y_pred = model.predict(X_test)
rmse = np.sqrt(mean_squared_error(y_test, y_pred))
print (rmse, '(rmse)', br)
print ('predict from new data:')
p1 = [X[244]]
p2 = [X[245], X[246]]
y1, y2 = model.predict(p1), model.predict(p2)
print (y[244], y1[0])
print (y[245], y2[0])
print (y[246], y2[1])
X_file = 'data/X_tips'
y_file = 'data/y_tips'
np.save(X_file, X)
np.save(y_file, y)
```

Go ahead and execute the code from Listing 4-1. Remember that you can find the example from the book's example download. You don't need to type the example by hand. It's easier to access the example download and copy/paste.

Your output from executing Listing 4-1 should resemble the following:

```
original data shape: (244, 7)

X and y shapes (post conversion):
(244, 12) (244,)

new X and y data point:
['30.0' 'NaN' '1' '0' '1' '0' '0' '0' '0' '1' '1' '0'] 4.5

new X and y data points:
['24.99' 'NaN' '0' '1' '0' '1' '1' '0' '0' '0' '0' '1'] 3.5
['19.99' 'NaN' '1' '0' '1' '0' '0' '0' '0' '1' '1' '0'] 2.0

new data shape: (247, 12)
```

```
new records post imputation (features and targets):
[30.          2.56967213  1.          0.          1.          0.
  0.          0.          0.          1.          1.          0.        ] 4.5
[24.99        2.56967213  0.          1.          0.          1.
  1.          0.          0.          0.          0.          1.        ] 3.5
[19.99        2.56967213  1.          0.          1.          0.
  0.          0.          0.          1.          1.          0.        ] 2.0

feature importance (first 6 features):
(0.7597845511444519, 'total_bill')
(0.0643775334380493, 'size')
(0.03663421916266647, 'non-smoker')
(0.033603977117169975, 'smoker')
(0.026410154999617023, 'sat')
(0.02186564474599064, 'sun')

<<LinearRegression>>

0.9474705746817206 (rmse)

predict from new data:
4.5 3.827512419066452
3.5 3.56649951075833
2.0 2.941595038244732
```

The code example begins by importing RandomForestRegressor, SimpleImputer, and mean_square_error as well as other requisite packages. The main block begins by loading tips data from a CSV file. It continues by feature engineering categorical variables with the pandas get_dummies function. As a reminder, *feature engineering* is using domain knowledge of the data set to create features that enable machine learning algorithms to work more effectively.

Pandas get_dummies one hot encodes by default. For example, sex is either *female* or *male*. With one-hot encoding, *female* becomes [1, 0] and *male* [0, 1]. Feature engineering must be conducted in this case because Scikit-Learn only works with numerical data. Next, feature set X and target y are created. Notice that the data shape now has twelve features as a result of one-hot encoding.

The next part of the code adds three new records to the data set. Also notice that we added a *NaN* feature, which means that the feature has no discernable value.

Scikit-Learn algorithms *cannot* handle *Nan* features, so we impute feature means with the SimpleImputer class. *Imputation* is the process of replacing missing data with substituted values. We next display the new records. Notice that all *Nan* values are replaced with their feature mean.

Tip   Imputation is a common technique for replacing missing data with substituted ones.

The last part of the code begins by displaying the six most important features with the help of *RandomForestRegressor*, which is a meta estimator that fits classifying decision trees on the data and uses averaging to improve performance and mitigate overfitting. Next, data is split into train-test subsets and trained with LinearRegression so we can calculate RMSE. *LinearRegression* models the relationship between features and the target. Finally, we make predictions from the new data where the 1st value is the actual target and the second is our prediction, and save X and y as NumPy files for future processing.

Tip   The goal of regression is to minimize (reduce) RMSE.

The next code example shown in Listing 4-2 uses DictVectorizer rather than get_dummies as an alternative option to one-hot encode categorical data.

Listing 4-2. Predicting from tips with DictVectorizer encoding

```python
import pandas as pd, numpy as np
from sklearn.feature_extraction import DictVectorizer
from sklearn.linear_model import LinearRegression
from sklearn.model_selection import train_test_split
from sklearn.metrics import mean_squared_error
from random import randint

if __name__ == "__main__":
    br = '\n'
    tips = pd.read_csv('data/tips.csv')
    data = tips.drop(['tip'], axis=1)
    target = tips['tip']
```

```
v = ['sex', 'smoker', 'day', 'time']
ls = data[v].to_dict(orient='records')
vector = DictVectorizer(sparse=False, dtype=int)
d = vector.fit_transform(ls)
print ('one hot encoding:')
print (d[0:3], br)
print ('encoding order:')
encode_order = vector.get_feature_names()
print (encode_order, br)
data = data.drop(['sex', 'smoker', 'day', 'time'], axis=1)
X = data.values
print ('feature shape after removing categorical columns:')
print (X.shape, br)
Xls, dls = X.tolist(), d.tolist()
X = [np.array(row + dls[i]) for i, row in enumerate(Xls)]
X = np.array(X)
y = target.values
print ('feature shape after adding encoded data back:')
print (X.shape, br)
X_train, X_test, y_train, y_test = train_test_split(
    X, y, random_state=0)
model = LinearRegression(fit_intercept=True)
model_name = model.__class__.__name__
print ('<<' + model_name + '>>', br)
model.fit(X_train, y_train)
y_pred = model.predict(X_test)
rmse = np.sqrt(mean_squared_error(y_test, y_pred))
print (rmse, '(rmse)', br)
print ('predict 1st test set element (actual/prediction):')
print (y_test[0], y_pred[0], br)
rints = [randint(0, y.shape[0]-1) for row in range(3)]
print ('random integers:', rints, br)
p = [X[rints[0]], X[rints[1]], X[rints[2]]]
y_p = model.predict(p)
y_p = list(np.around(y_p, 2))
```

```
print (y_p, '(predicted)')
print ([y[rints[0]], y[rints[1]], y[rints[2]]], '(actual)')
```

Your output from executing Listing 4-2 should resemble the following:

```
One hot encoding:
[[0 0 1 0 1 0 1 0 1 0]
 [0 0 1 0 0 1 1 0 1 0]
 [0 0 1 0 0 1 1 0 1 0]]
```

```
encoding order:
['day=Fri', 'day=Sat', 'day=Sun', 'day=Thur', 'sex=Female', 'sex=Male',
'smoker=No','smoker=Yes', 'time=Dinner', 'time=Lunch']
```

```
feature shape after removing categorical columns:
(244, 2)
```

```
feature shape after adding encoded data back:
(244, 12)
```

```
<<LinearRegression>>
```

```
0.9636287548943022 (rmse)
```

```
predict 1st test set element (actual/prediction):
2.64 2.8121130438023094
```

```
random integers: [202, 13, 143]
```

```
[2.19, 3.21, 4.52] (predicted)
[2.0, 3.0, 5.0] (actual)
```

The code begins by importing DictVectorizer as well as other requisite packages. The main block loads tips data, strips away feature tip and places the remaining features in variable *data*, and places feature tip in variable *target*. Next, we indicate features that need encoding by placing them in variable *v*.

The code continues by stripping away only features that need encoding and placing the result in variable *ls*. A DictVectorizer instance is then created and placed in variable *vector*. DictVectorizer is a Scikit-Learn technique that transforms lists of feature-value mappings to vectors.

Data in *vector* is the then fitted (or trained) and transformed to one-hot encoded values, and the result is placed in variable *d*. To see encoding order, use function get_feature_names. Next, encoded features are dropped from variable *data* so we can start building feature set X. Whew!

The code continues by creating a list based on X and another based on *d* so we can concatenate X with one-hot encoded values contained in *d*. All that remains is to convert X and y to NumPy values. The code ends by splitting X and y into train-test subsets, training with LinearRegression, calculating RMSE, and making predictions. RMSE is a bit higher because we didn't add new data like in the previous example.

Tip   Use get_dummies for one-hot encoding in most instances.

Although DictVectorizer is more difficult to implement than get_dummies, it has the advantage of being sparse. That is, absent features need not be stored. In addition, DictVectorizer is a useful representation transformation for training sequence classifiers in Natural Language Processing models that typically work by extracting feature windows around a particular word of interest.

The final code example in this section shown in Listing 4-3 loads engineered data (from NumPy files) into X and y, and calculates RMSE for several regression algorithms that implement regularization. *Regularization* is a technique used to reduce error by fitting a function (or algorithm) appropriately on a given data set to mitigate overfitting.

Listing 4-3. Predicting from tips with regression regularization models

```
import numpy as np
from sklearn.linear_model import LinearRegression, Ridge,\
     Lasso, ElasticNet, SGDRegressor
from sklearn.model_selection import train_test_split
from sklearn.metrics import mean_squared_error
from sklearn.preprocessing import StandardScaler

def get_scores(model, Xtest, ytest):
    y_pred = model.predict(Xtest)
    return np.sqrt(mean_squared_error(ytest, y_pred)),\
        model.__class__.__name__
```

```python
if __name__ == "__main__":
    br = '\n'
    X = np.load('data/X_tips.npy')
    y = np.load('data/y_tips.npy')
    X_train, X_test, y_train, y_test = train_test_split(
        X, y, random_state=0)
    print ('rmse:')
    lr = LinearRegression().fit(X_train, y_train)
    rmse, lr_name = get_scores(lr, X_test, y_test)
    print (rmse, '(' + lr_name + ')')
    rr = Ridge(random_state=0).fit(X_train, y_train)
    rmse, rr_name = get_scores(rr, X_test, y_test)
    print (rmse, '(' + rr_name + ')')
    lasso = Lasso(random_state=0).fit(X_train, y_train)
    rmse, lasso_name = get_scores(lasso, X_test, y_test)
    print (rmse, '(' + lasso_name + ')')
    en = ElasticNet(random_state=0).fit(X_train, y_train)
    rmse, en_name = get_scores(en, X_test, y_test)
    print (rmse, '(' + en_name + ')')
    sgdr = SGDRegressor(random_state=0, max_iter=1000, tol=0.001)
    sgdr.fit(X_train, y_train)
    rmse, sgdr_name = get_scores(sgdr, X_test, y_test)
    print (rmse, '(' + sgdr_name + ')', br)
    scaler = StandardScaler()
    X_train_std = scaler.fit_transform(X_train)
    X_test_std = scaler.fit_transform(X_test)
    print ('rmse std:')
    lr_std = LinearRegression().fit(X_train_std, y_train)
    rmse, lr_name = get_scores(lr_std, X_test_std, y_test)
    print (rmse, '(' + lr_name + ')')
    rr_std = Ridge(random_state=0).fit(X_train_std, y_train)
    rmse, rr_name = get_scores(rr_std, X_test_std, y_test)
    print (rmse, '(' + rr_name + ')')
    lasso_std = Lasso(random_state=0).fit(X_train_std, y_train)
```

```
rmse, lasso_name = get_scores(lasso_std, X_test_std, y_test)
print (rmse, '(' + lasso_name + ')')
en_std = ElasticNet(random_state=0)
en_std.fit(X_train_std, y_train)
rmse, en_name = get_scores(en_std, X_test_std, y_test)
print (rmse, '(' + en_name + ')')
sgdr_std = SGDRegressor(random_state=0, max_iter=1000, tol=0.001)
sgdr_std.fit(X_train_std, y_train)
rmse, sgdr_name = get_scores(sgdr_std, X_test_std, y_test)
print (rmse, '(' + sgdr_name + ')')
```

Your output from executing Listing 4-3 should resemble the following:

```
rmse:
0.9474705746817206 (LinearRegression)
0.9469115898683899 (Ridge)
0.9439950256305224 (Lasso)
0.9307377813721578 (ElasticNet)
1.7005504977258326 (SGDRegressor)

rmse std:
0.9007751177881488 (LinearRegression)
0.9014055340745654 (Ridge)
1.333812899498391 (Lasso)
1.1310151423347359 (ElasticNet)
0.9021020134681715 (SGDRegressor)
```

The code example begins by importing Ridge, Lasso, ElasticNet, SGDRegressor, and other requisite packages. Function get_scores returns RMSE. The main block begins by loading data from NumPy files into X and y. Next, data is split into train-test subsets.

The remainder of the code trains data with LinearRegression and several regression models that implement regularization. Ridge, Lasso, ElasticNet, and SGDRegressor are popular Scikit-Learn regression algorithms introduced to regularize LinearRegression. Regularization reduces error by fitting a model appropriately on a given train set to mitigate overfitting. That is, regularization discourages learning a more complex model to mitigate the risk of overfitting.

Tip   Use regularization to reduce error and minimize overfitting with regression models.

Ridge regression imposes a penalty on the size of the coefficients. *Lasso* regression derives solutions with fewer parameter values (or a sparse model) that effectively reduces the number of variables upon which the solution is dependent.

Lasso uses L1 regularization and Ridge uses L2 regularization.

The key difference between L1 and L2 is the penalty term. Ridge adds *squared magnitude* of coefficient as penalty to the loss function to mitigate overfitting. Lasso (Least Absolute Shrinkage and Selection Operator) adds *absolute value of magnitude* of coefficient as penalty to the loss function, which works well when we have a huge number of features.

A loss (cost) function is one that maps an event (or values) of one or more features onto a real number that represents a cost associated with the event. The main difference between techniques is that Lasso shrinks the less important features' coefficient to zero, which effectively removes them from consideration. Lasso provides the same results for dense and sparse data, and with sparse data the speed is improved.

ElasticNet regression embraces both L1 and L2 penalties as the regularizer. Combining L1 and L2 allows for learning a sparse model where few of the weights are nonzero like Lasso while still maintaining the regularization properties of Ridge. ElasticNet is at its best with multiple features that are correlated. A practical advantage of leveraging the tradeoff between Lasso and Ridge is that it allows ElasticNet to inherit some of Ridge's stability under rotation while still performing well with a sparse model.

SGDRegressor performs by minimizing a regularized empirical loss with stochastic gradient descent (SGD). That is, the gradient of the loss is estimated with each sample, and the model is then updated along the way with a decreasing strength schedule (or learning rate). The regularizer is a penalty added to the loss function that shrinks parameters toward zero using either L1 (Lasso) or L2 (Ridge), or a combination of both (ElasticNet).

So, choice of regularization technique is highly dependent on the nature of the data. Sparse data would indicate beginning with Lasso. Creating overly complex models would indicate turning to Ridge. ElasticNet offers a compromise. SGDRegressor attempts to do it all. With relatively small data sets, trying all of these techniques offers no problems. But, large data sets are a different story because of the enormous processing time required (or high computational expense).

Notice that sometimes scaling (or standardization) improves performance and sometimes it doesn't. For instance, scaling helped with LinearRegression, Ridge, and SGDRegressor but hurt with Lasso and ElasticNet.

Tip Experimentation is an excellent way to improve performance if you have the time, patience, and computing resources.

Regressing boston

The first code example shown in Listing 4-4 displays feature importance from the boston data set, trains with RandomForestRegressor, and calculates RMSE with and without noise.

Listing 4-4. Exploring boston data with RandomForestRegressor

```
import numpy as np, pandas as pd
from sklearn.datasets import load_boston
from sklearn.model_selection import train_test_split
from sklearn.ensemble import RandomForestRegressor
from sklearn.metrics import mean_squared_error

if __name__ == "__main__":
    br = '\n'
    boston = load_boston()
    X = boston.data
    y = boston.target
    print ('feature shape', X.shape)
    print ('target shape', y.shape, br)
    keys = boston.keys()
    rfr = RandomForestRegressor(random_state=0, n_estimators=100)
    rfr.fit(X, y)
    features = boston.feature_names
    feature_importances = rfr.feature_importances_
    importance = sorted(zip(feature_importances, features), reverse=True)
```

```python
[print (row) for row in importance]
print ()
X_train, X_test, y_train, y_test = train_test_split(
    X, y, random_state=0)
rfr = RandomForestRegressor(random_state=0, n_estimators=100)
rfr.fit(X_train, y_train)
rfr_name = rfr.__class__.__name__
y_pred = rfr.predict(X_test)
rmse = np.sqrt(mean_squared_error(y_test, y_pred))
print (rfr_name + ' (rmse):', rmse, br)
cols = list(features) + ['target']
data = pd.DataFrame(data=np.c_[X, y], columns=cols)
print ('boston dataset sample:')
print (data[['RM', 'LSTAT', 'DIS', 'CRIM', 'NOX', 'PTRATIO', 'target']].
            head(3), br)
print ('data set before removing noise:', data.shape)
noise = data.loc[data['target'] >= 50]
data = data.drop(noise.index)
print ('data set without noise:', data.shape, br)
X = data.loc[:, data.columns != 'target'].values
y = data['target'].values
print ('cleansed feature shape:', X.shape)
print ('cleansed target shape:', y.shape, br)
X_train, X_test, y_train, y_test = train_test_split(
    X, y, random_state=0)
rfr = RandomForestRegressor(random_state=0, n_estimators=100)
rfr.fit(X_train, y_train)
y_pred = rfr.predict(X_test)
rmse = np.sqrt(mean_squared_error(y_test, y_pred))
print (rfr_name + ' (rmse):', rmse)
X_file = 'data/X_boston'
y_file = 'data/y_boston'
np.save(X_file, X)
np.save(y_file, y)
```

Your output from executing Listing 4-4 should resemble the following:

```
feature shape (506, 13)
target shape (506,)

(0.45730362625767496, 'RM')
(0.35008661885681375, 'LSTAT')
(0.06518862820215894, 'DIS')
(0.040989617257001, 'CRIM')
(0.02024797563034355, 'NOX')
(0.015576365835498516, 'PTRATIO')
(0.015524054184831321, 'TAX')
(0.011764308556043926, 'AGE')
(0.011324966974602932, 'B')
(0.005912139937999768, 'INDUS')
(0.003916064249793193, 'RAD')
(0.0011173446269339175, 'ZN')
(0.0010482894303040916, 'CHAS')

RandomForestRegressor (rmse): 4.091149842219918

boston dataset sample:
      RM  LSTAT    DIS     CRIM    NOX  PTRATIO  target
0  6.575   4.98  4.0900  0.00632  0.538    15.3    24.0
1  6.421   9.14  4.9671  0.02731  0.469    17.8    21.6
2  7.185   4.03  4.9671  0.02729  0.469    17.8    34.7

data set before removing noise: (506, 14)
data set without noise: (490, 14)

cleansed feature shape: (490, 13)
cleansed target shape: (490,)

RandomForestRegressor (rmse): 3.37169151536684
```

The code example begins by importing requisite packages. The main block loads boston data from sklearn.datasets into X and y and displays the shape. Next, RandomForestRegressor trains on the full data set (X and y) to create and display feature importance. The code continues by splitting data into train-test subsets and training (X_train, y_train) with RandomForestRegressor. RMSE is calculated and displayed.

Random Forest is an ensemble technique capable of performing both regression and classification by using multiple decision trees and bagging. Bagging involves training each decision tree on a different data sample with replacement. The idea is to combine multiple decision trees to determine the result rather than relying on individual decision trees.

The code continues by reading X and y into a Pandas DataFrame and displaying the first three records. The noise is then removed from the DataFrame and saved into X and y. The cleansed data is split into train-test subsets and trained with RandomForestRegressor. Notice that RMSE is quite a bit lower (less error) with the cleansed data. The code concludes by saving X and y as NumPy files.

Sixteen data points have an *MEDV* value of 50.0, which likely contain missing or censored values and can be considered noise. So, we removed them from consideration. For more information on noise in the boston data set, consult the following link: *https://www.ritchieng.com/machine-learning-project-boston-home-prices/*.

The final code example in this section shown in Listing 4-5 loads the cleansed (noise removed) boston data and calculates RMSE with LinearRegression, regularization models, and RandomForestRegressor.

Listing 4-5. Exploring boston data with regression algorithms

```
import numpy as np
from sklearn.datasets import load_boston
from sklearn.model_selection import train_test_split
from sklearn.linear_model import LinearRegression, Ridge,\
     Lasso, ElasticNet, SGDRegressor
from sklearn.ensemble import RandomForestRegressor
from sklearn.metrics import mean_squared_error
from sklearn.preprocessing import StandardScaler

def get_scores(model, Xtest, ytest):
    y_pred = model.predict(Xtest)
    return np.sqrt(mean_squared_error(ytest, y_pred)),\
           model.__class__.__name__

if __name__ == "__main__":
    br = '\n'
    X = np.load('data/X_boston.npy')
```

```python
y = np.load('data/y_boston.npy')
print ('feature shape', X.shape)
print ('target shape', y.shape, br)
X_train, X_test, y_train, y_test = train_test_split( X, y, random_state=0)
print ('rmse:')
rfr = RandomForestRegressor(random_state=0, n_estimators=100)
rfr.fit(X_train, y_train)
rmse, rfr_name = get_scores(rfr, X_test, y_test)
print (rmse, '(' + rfr_name + ')')
lr = LinearRegression().fit(X_train, y_train)
rmse, lr_name = get_scores(lr, X_test, y_test)
print (rmse, '(' + lr_name + ')')
ridge = Ridge(random_state=0).fit(X_train, y_train)
rmse, ridge_name = get_scores(ridge, X_test, y_test)
print (rmse, '(' + ridge_name + ')')
lasso = Lasso(random_state=0).fit(X_train, y_train)
rmse, lasso_name = get_scores(lasso, X_test, y_test)
print (rmse, '(' + lasso_name + ')')
en = ElasticNet(random_state=0).fit(X_train, y_train)
rmse, en_name = get_scores(en, X_test, y_test)
print (rmse, '(' + en_name + ')')
scaler = StandardScaler()
X_train_std = scaler.fit_transform(X_train)
X_test_std = scaler.fit_transform(X_test)
sgdr_std = SGDRegressor(random_state=0, max_iter=1000, tol=0.001)
sgdr_std.fit(X_train_std, y_train)
rmse, sgdr_name = get_scores(sgdr_std, X_test_std, y_test)
print (rmse, '(' + sgdr_name + ' - scaled)')
```

Your output from executing Listing 4-5 should resemble the following:

```
feature shape (490, 13)
target shape (490,)

rmse:
3.37169151536684 (RandomForestRegressor)
4.236710574387242 (LinearRegression)
4.2526986026173486 (Ridge)
5.097231463859832 (Lasso)
4.88844846745213 (ElasticNet)
4.410035683951274 (SGDRegressor - scaled)
```

The code begins by importing requisite packages. Function get_scores returns RMSE and algorithm name. The main block begins by loading data from NumPy files and splitting it into train-test subsets. Data is then trained with RandomForestRegressor, LinearRegression, Ridge, Lasso, ElasticNet, and SGDRegressor. Notice that RandomForestRegressor outperformed all of the regularization algorithms as its RMSE is the lowest.

Regressing wine data

The first code example shown in Listing 4-6 loads red wine data from a CSV file, displays feature importance, and saves data to NumPy files.

Listing 4-6. Exploring and saving red wine data

```
import numpy as np, pandas as pd
from sklearn.ensemble import RandomForestRegressor

if __name__ == "__main__":
    br = '\n'
    f = 'data/redwine.csv'
    red_wine = pd.read_csv(f)
    X = red_wine.drop(['quality'], axis=1)
    y = red_wine['quality']
    print (X.shape)
```

```
print (y.shape, br)
features = list(X)
rfr = RandomForestRegressor(random_state=0, n_estimators=100)
rfr.fit(X, y)
feature_importances = rfr.feature_importances_
importance = sorted(zip(feature_importances, features), reverse=True)
for row in importance:
    print (row)
print ()
print (red_wine[['alcohol', 'sulphates', 'volatile acidity',
                'total sulfur dioxide', 'quality']]. head())
X_file = 'data/X_red'
y_file = 'data/y_red'
np.save(X_file, X)
np.save(y_file, y)
```

Your output from executing Listing 4-6 should resemble the following:

```
(1599, 11)
(1599,)

(0.27432500255956216, 'alcohol')
(0.13700073893077233, 'sulphates')
(0.13053941311188708, 'volatile acidity')
(0.08068199773601588, 'total sulfur dioxide')
(0.06294612644261727, 'chlorides')
(0.057730976351602854, 'pH')
(0.055499749756166, 'residual sugar')
(0.05198192402458334, 'density')
(0.05114079873500658, 'fixed acidity')
(0.049730883807319035, 'free sulfur dioxide')
(0.04842238854446754, 'citric acid')
```

	alcohol	sulphates	volatile acidity	total sulfur dioxide	quality
0	9.4	0.56	0.70	34.0	5.0
1	9.8	0.68	0.88	67.0	5.0
2	9.8	0.65	0.76	54.0	5.0
3	9.8	0.58	0.28	60.0	6.0
4	9.4	0.56	0.70	34.0	5.0

The code begins by importing requisite packages. The main block loads red wine data from a CSV file. Next, feature set X is created by stripping off the target column quality from the Pandas DataFrame, and then target y is created from the quality column. X and y shapes are then displayed. The code concludes by displaying feature importance with the help of RandomForestRegressor and saving data to NumPy files.

The next code example shown in Listing 4-7 experiments with red wine data using a variety of regression algorithms.

Listing 4-7. Exploring red wine data with regression algorithms

```
import numpy as np, pandas as pd
from sklearn.ensemble import RandomForestRegressor
from sklearn.model_selection import train_test_split
from sklearn.linear_model import LinearRegression,\
     Ridge, Lasso, ElasticNet, SGDRegressor
from sklearn.metrics import mean_squared_error
from sklearn.preprocessing import StandardScaler
from sklearn.preprocessing import PolynomialFeatures
from sklearn.pipeline import Pipeline
import matplotlib.pyplot as plt, seaborn as sns

def get_scores(model, Xtest, ytest):
    y_pred = model.predict(Xtest)
    return np.sqrt(mean_squared_error(ytest, y_pred)),\
        model.__class__.__name__

if __name__ == "__main__":
    br = '\n'
    d = dict()
    X = np.load('data/X_red.npy')
    y = np.load('data/y_red.npy')
```

```
X_train, X_test, y_train, y_test = train_test_split(
    X, y, test_size=0.2, random_state=0)
print ('rmse (unscaled):')
rfr = RandomForestRegressor(random_state=0, n_estimators=100)
rfr.fit(X_train, y_train)
rmse, rfr_name = get_scores(rfr, X_test, y_test)
d['rfr'] = [rmse]
print (rmse, '(' + rfr_name + ')')
lr = LinearRegression().fit(X_train, y_train)
rmse, lr_name = get_scores(lr, X_test, y_test)
d['lr'] = [rmse]
print (rmse, '(' + lr_name + ')')
ridge = Ridge(random_state=0).fit(X_train, y_train)
rmse, ridge_name = get_scores(ridge, X_test, y_test)
d['ridge'] = [rmse]
print (rmse, '(' + ridge_name + ')')
lasso = Lasso(random_state=0).fit(X_train, y_train)
rmse, lasso_name = get_scores(lasso, X_test, y_test)
d['lasso'] = [rmse]
print (rmse, '(' + lasso_name + ')')
en = ElasticNet(random_state=0).fit(X_train, y_train)
rmse, en_name = get_scores(en, X_test, y_test)
d['en'] = [rmse]
print (rmse, '(' + en_name + ')')
sgdr = SGDRegressor(random_state=0, max_iter=1000, tol=0.001)
sgdr.fit(X_train, y_train)
rmse, sgdr_name = get_scores(sgdr, X_test, y_test)
print (rmse, '(' + sgdr_name + ')', br)
scaler = StandardScaler()
X_train_std = scaler.fit_transform(X_train)
X_test_std = scaler.fit_transform(X_test)
print ('rmse scaled:')
lr_std = LinearRegression().fit(X_train_std, y_train)
```

```
rmse, lr_std_name = get_scores(lr_std, X_test_std, y_test)
print (rmse, '(' + lr_std_name + ')')
rr_std = Ridge(random_state=0).fit(X_train_std, y_train)
rmse, rr_std_name = get_scores(rr_std, X_test_std, y_test)
print (rmse, '(' + rr_std_name + ')')
lasso_std = Lasso(random_state=0).fit(X_train_std, y_train)
rmse, lasso_std_name = get_scores(lasso_std, X_test_std, y_test)
print (rmse, '(' + lasso_std_name + ')')
en_std = ElasticNet(random_state=0).fit(X_train_std, y_train)
rmse, en_std_name = get_scores(en_std, X_test_std, y_test)
print (rmse, '(' + en_std_name + ')')
sgdr_std = SGDRegressor(random_state=0, max_iter=1000, tol=0.001)
sgdr_std.fit(X_train_std, y_train)
rmse, sgdr_std_name = get_scores(sgdr_std, X_test_std, y_test)
d['sgdr_std'] = [rmse]
print (rmse, '(' + sgdr_std_name + ')', br)
pipe = Pipeline([('poly', PolynomialFeatures(degree=2)),
                ('linear', LinearRegression())])
model = pipe.fit(X_train, y_train)
rmse, poly_name = get_scores(model, X_test, y_test)
d['poly'] = [rmse]
print (PolynomialFeatures().__class__.__name__, '(rmse):')
print (rmse, '(' + poly_name + ')')
algo, rmse = [], []
for key, value in d.items():
    algo.append(key)
    rmse.append(value[0])
plt.figure('RMSE')
sns.set(style="whitegrid")
ax = sns.barplot(algo, rmse)
plt.title('Red Wine Algorithm Comparison')
```

```
plt.xlabel('regressor')
plt.ylabel('RMSE')
plt.show()
```

Your output from executing Listing 4-7 should resemble the following:

```
rmse (unscaled):
0.5694654840286635 (RandomForestRegressor)
0.6200574149384266 (LinearRegression)
0.6185762657415644 (Ridge)
0.7455442007369433 (Lasso)
0.7450232657227877 (ElasticNet)
51120537008.37402 (SGDRegressor)

rmse scaled:
0.6216027053463463 (LinearRegression)
0.6215826846730879 (Ridge)
0.7584549718351333 (Lasso)
0.7584549718351333 (ElasticNet)
0.6234205584462227 (SGDRegressor)

PolynomialFeatures (rmse):
0.6382400985644077 (Pipeline)
```

Listing 4-7 also displays Figure 4-1. Figure 4-1 provides a visualization of RMSE scores for the algorithms used in this experiment.

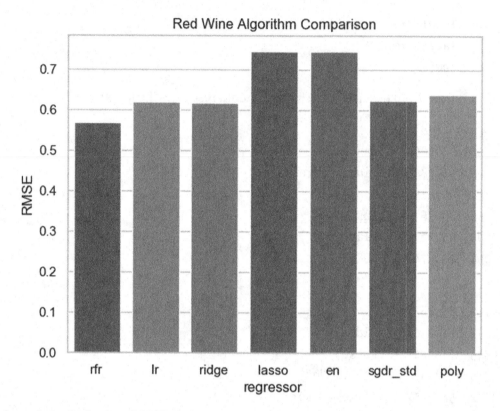

Figure 4-1. *Red wine RMSE score comparison*

The code begins by importing PolynomialFeatures, Pipeline, seaborn, and other requisite packages. Function get_scores returns RMSE and model name. The main block begins by creating a dictionary to store the best RMSE scores from the training experiments. Algorithms train data with and without scaling, and the best score is saved in dictionary *d*. Next, data is split into train-test subsets.

The code continues by training *unscaled* data with LinearRegression, Ridge, Lasso, ElasticNet, SGDRegressor, and RandomForestRegressor. The best performing algorithm on this data set is RandomForestRegressor with a RMSE of approximately 0.569. The code then trains *scaled* data with LinearRegression, Ridge, Lasso, ElasticNet, and SGDRegressor. The best performing algorithm on this data set is Ridge with a RMSE of approximately 0.622, which is not better than its unscaled RMSE. Although RMSE SGDRegressor was not the best performer, notice how much scaling impacts the algorithm! Finally, unscaled data is trained with PolynomialFeatures.

PolynomialFeatures offers an opportunity to improve performance by transforming inputs rather than improving a model. Polynomial regression allows for a linear combination of an input raised to varying degrees. In this example, we square inputs (degree=2) to explore impact on performance. To train data, we pipe PolynomialFeatures to LinearRegression.

We can experiment with degree to see what happens to performance. We can cube inputs (degree=3), quadruple inputs (degree=4), and so on. Polynomial models can be very useful for nonlinear machine learning experiments, but be careful with high-order polynomial models because they generally are not well-behaved. That is, they can produce dramatic unwanted fluctuations. Regularization can mitigate polynomial misbehavior.

Training with PolynomialFeatures is accomplished by piping the transformed input (squaring data) into LinearRegression. The code concludes by creating a visualization based on the best RMSE scores stored in dictionary *d*.

The next code example shown in Listing 4-8 experiments with polynomial fitting. In the previous example, we squared input data, trained the model, and calculated RMSE. In this one, we take input data to the second (squared), third, and fourth power, train each model, and calculate and display RMSE for comparison. Another change from the previous code is instead of piping PolynomialFeatures into LinearRegression, we transform train and test data with the PolynomialFeatures algorithm. We then train LinearRegression with the transformed data. Once data is trained, we display RMSE for each experiment.

Listing 4-8. Polynomial fitting with red wine data

```
import numpy as np, pandas as pd
from sklearn.model_selection import train_test_split
from sklearn.linear_model import LinearRegression
from sklearn.metrics import mean_squared_error
from sklearn.preprocessing import PolynomialFeatures

def get_scores(model, Xtest, ytest):
    y_pred = model.predict(Xtest)
    return np.sqrt(mean_squared_error(ytest, y_pred)),\
           model.__class__.__name__

if __name__ == "__main__":
    br = '\n'
    d = dict()
```

```
X = np.load('data/X_red.npy')
y = np.load('data/y_red.npy')
X_train, X_test, y_train, y_test =  train_test_split(
    X, y, test_size=0.2, random_state=0)
poly = PolynomialFeatures(degree=2)
poly.fit(X_train, y_train)
X_train_poly = poly.transform(X_train)
lr = LinearRegression().fit(X_train_poly, y_train)
X_test_poly = poly.transform(X_test)
rmse, lr_name = get_scores(lr, X_test_poly, y_test)
print (rmse, '(squared polynomial fitting)')
poly = PolynomialFeatures(degree=3)
poly.fit(X_train, y_train)
X_train_poly = poly.transform(X_train)
lr = LinearRegression().fit(X_train_poly, y_train)
X_test_poly = poly.transform(X_test)
rmse, lr_name = get_scores(lr, X_test_poly, y_test)
print (rmse, '(cubic polynomial fitting)')
poly = PolynomialFeatures(degree=4)
poly.fit(X_train, y_train)
X_train_poly = poly.transform(X_train)
lr = LinearRegression().fit(X_train_poly, y_train)
X_test_poly = poly.transform(X_test)
rmse, lr_name = get_scores(lr, X_test_poly, y_test)
print (rmse, '(quartic polynomial fitting)')
```

Your output from executing Listing 4-8 should resemble the following:

```
0.6382400985644077 (squared polynomial fitting)
0.8284645679714848 (cubic polynomial fitting)
97.85391125320886 (quartic polynomial fitting)
```

The code imports requisite packages. Function get_scores returns RMSE and model name. The main block loads data into X and y. It continues by splitting data into train-test subsets. Next, train and test data is fitted with PolynomialFeatures with squared input data (degree=2). Train data is then transformed. LinearRegression trains on the transformed data and RMSE is displayed. The same process is followed with degree=3 and degree=4. For this data set, squaring the input provides the best RMSE.

Tip PolynomialFeatures can be a very useful technique for modeling nonlinear data sets, and it is easy to implement.

The next code example shown in Listing 4-9 loads white wine data from a CSV file, displays feature importance, and saves data to NumPy files.

Listing 4-9. Exploring and saving white wine data

```python
import numpy as np, pandas as pd
from sklearn.ensemble import RandomForestRegressor

if __name__ == "__main__":
    br = '\n'
    f = 'data/whitewine.csv'
    white_wine = pd.read_csv(f)
    X = white_wine.drop(['quality'], axis=1)
    y = white_wine['quality']
    print (X.shape)
    print (y.shape, br)
    features = list(X)
    rfr = RandomForestRegressor(random_state=0, n_estimators=100)
    rfr.fit(X, y)
    feature_importances = rfr.feature_importances_
    importance = sorted(zip(feature_importances, features), reverse=True)
    for row in importance:
        print (row)
    print ()
    print (white_wine[['alcohol', 'sulphates', 'volatile acidity',
                       'total sulfur dioxide', 'quality']]. head())
    X_file = 'data/X_white'
    y_file = 'data/y_white'
    np.save(X_file, X)
    np.save(y_file, y)
```

131

Your output from executing Listing 4-9 should resemble the following:

```
(4898, 11)
(4898,)

(0.24186185906056268, 'alcohol')
(0.1251626059551235, 'volatile acidity')
(0.11524332271725685, 'free sulfur dioxide')
(0.07170261049200727, 'pH')
(0.06940456299270928, 'total sulfur dioxide')
(0.06899334812486085, 'residual sugar')
(0.06259740092261244, 'chlorides')
(0.06227404207074219, 'sulphates')
(0.061557623671947746, 'density')
(0.060982526101159625, 'citric acid')
(0.060220097891017656, 'fixed acidity')
```

	alcohol	sulphates	volatile acidity	total sulfur dioxide	quality
0	8.8	0.45	0.27	170.0	6.0
1	9.5	0.49	0.30	132.0	6.0
2	10.1	0.44	0.28	97.0	6.0
3	9.9	0.40	0.23	186.0	6.0
4	9.9	0.40	0.23	186.0	6.0

The code begins by importing requisite packages. The main block loads white wine data from a CSV file. Next, feature set X is created by stripping off the target column quality from the Pandas DataFrame, and target y is created from the quality column. X and y shapes are then displayed. Notice that the white wine data set is composed of 4898 data elements while red wine data was much smaller at 1599. The code concludes by displaying feature importance with the help of RandomForestRegressor and saving data to NumPy files.

The final code example shown in Listing 4-10 experiments with white wine data using a variety of regression algorithms.

Listing 4-10. Exploring white wine data with regression algorithms

```
import numpy as np, pandas as pd
from sklearn.ensemble import RandomForestRegressor
from sklearn.model_selection import train_test_split
from sklearn.linear_model import LinearRegression,\
    Ridge, Lasso, ElasticNet, SGDRegressor
from sklearn.metrics import mean_squared_error
from sklearn.preprocessing import StandardScaler
from sklearn.preprocessing import PolynomialFeatures
from sklearn.pipeline import Pipeline
import matplotlib.pyplot as plt, seaborn as sns

def get_scores(model, Xtest, ytest):
    y_pred = model.predict(Xtest)
    return np.sqrt(mean_squared_error(ytest, y_pred)),\
        model.__class__.__name__

if __name__ == "__main__":
    br = '\n'
    d = dict()
    X = np.load('data/X_white.npy')
    y = np.load('data/y_white.npy')
    X_train, X_test, y_train, y_test =  train_test_split(
        X, y, test_size=0.2, random_state=0)
    print ('rmse (unscaled):')
    rfr = RandomForestRegressor(random_state=0, n_estimators=100)
    rfr.fit(X_train, y_train)
    rmse, rfr_name = get_scores(rfr, X_test, y_test)
    d['rfr'] = [rmse]
    print (rmse, '(' + rfr_name + ')')
    lr = LinearRegression().fit(X_train, y_train)
    rmse, lr_name = get_scores(lr, X_test, y_test)
    d['lr'] = [rmse]
    print (rmse, '(' + lr_name + ')')
    ridge = Ridge(random_state=0).fit(X_train, y_train)
```

```
rmse, ridge_name = get_scores(ridge, X_test, y_test)
d['ridge'] = [rmse]
print (rmse, '(' + ridge_name + ')')
lasso = Lasso(random_state=0).fit(X_train, y_train)
rmse, lasso_name = get_scores(lasso, X_test, y_test)
d['lasso'] = [rmse]
print (rmse, '(' + lasso_name + ')')
en = ElasticNet(random_state=0).fit(X_train, y_train)
rmse, en_name = get_scores(en, X_test, y_test)
d['en'] = [rmse]
print (rmse, '(' + en_name + ')', br)
scaler = StandardScaler()
X_train_std = scaler.fit_transform(X_train)
X_test_std = scaler.fit_transform(X_test)
print ('rmse scaled:')
sgd = SGDRegressor(max_iter=1000, tol=0.001, random_state=0)
sgd.fit(X_train_std, y_train)
rmse, sgd_name = get_scores(sgd, X_test_std, y_test)
d['sgd'] = [rmse]
print (rmse, '(' + sgd_name + ')', br)
pipe = Pipeline([('poly', PolynomialFeatures(degree=2)),
                ('linear', LinearRegression())])
model = pipe.fit(X_train, y_train)
rmse, pf_name = get_scores(model, X_test, y_test)
d['poly'] = [rmse]
print (PolynomialFeatures().__class__.__name__,'(rmse):')
print (rmse, '(' + pf_name + ')')
poly = PolynomialFeatures(degree=2)
poly.fit(X_train, y_train)
X_train_poly = poly.transform(X_train)
lr = LinearRegression().fit(X_train_poly, y_train)
X_test_poly = poly.transform(X_test)
rmse, lr_name = get_scores(lr, X_test_poly, y_test)
```

```
print (rmse, '(without Pipeline)')
algo, rmse = [], []
for key, value in d.items():
    algo.append(key)
    rmse.append(value[0])
plt.figure('RMSE')
sns.set(style="whitegrid")
ax = sns.barplot(algo, rmse)
plt.title('White Wine Algorithm Comparison')
plt.xlabel('regressor')
plt.ylabel('RMSE')
plt.show()
```

Your output from executing Listing 4-10 should resemble the following:

```
rmse (unscaled):
0.687111151629689 (RandomForestRegressor)
0.8123086554972433 (LinearRegression)
0.8141615403447382 (Ridge)
0.9255803421282806 (Lasso)
0.9242810596011943 (ElasticNet)

rmse scaled:
0.8092835779827245 (SGDRegressor)

PolynomialFeatures (rmse):
0.7767527802246017 (Pipeline)
0.7767527802246017 (without Pipeline)
```

Listing 4-10 also displays Figure 4-2. Figure 4-2 provides a visualization of RMSE scores for the algorithms used in this experiment.

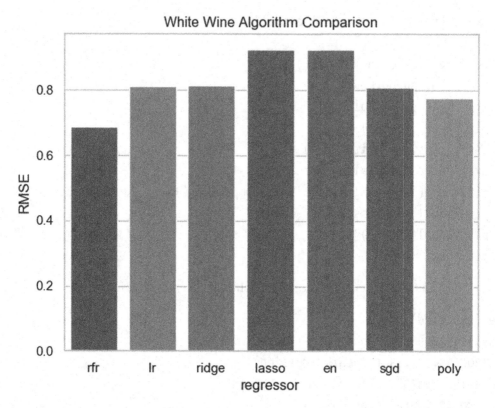

Figure 4-2. *White wine RMSE score comparison*

The code begins by importing requisite packages. Function get_scores returns RMSE and model name. The main block begins by creating a dictionary to store the best RMSE scores from the training experiments. It continues by loading white wine data into X and y from NumPy files. Data is then split into train-test subsets. Next, algorithms train data with and without scaling, and the best score is saved in dictionary *d*.

The code continues by training *unscaled* data with LinearRegression, Ridge, Lasso, ElasticNet, SGDRegressor, and RandomForestRegressor. The best performing algorithm on this data set is RandomForestRegressor with a RMSE of approximately 0.687. The code then trains *scaled* data with SGDRegressor. I had already determined (through experimentation not shown here) that RMSE for Ridge, Lasso, and ElasticNet was not improved with scaling, so I didn't include the code. Finally, unscaled data is trained with PolynomialFeatures with and without Pipeline. Since both RMSE scores are identical, it shouldn't matter which technique is used.

CHAPTER 5

Scikit-Learn Classifier Tuning from Simple Training Sets

Tuning is the process of maximizing an algorithm's performance without overfitting, underfitting, or creating high variance. *Overfitting* is when an algorithm trains data so exactly that it may fail to fit new data or predict future results reliably. Overfitting usually occurs when a model is too complex for the data it is trying to train. An overly complex model trains data very well, but also fits noise that is not a part of the data. So, when used to train new data, noise is introduced causing unpredictable results.

Underfitting is when an algorithm cannot adequately capture the underlying structure of the data. An underfitting model underperforms because it is not complex enough to capture the meaning of the data. *High variance* is when an algorithm introduces too much error into prediction resulting in poor performance.

High-performance tuning is accomplished by selecting optimal hyperparameters from a model. A model *hyperparameter* is a configuration external to the model and whose value cannot be estimated from data. Most machine learning algorithms have a set of hyperparameters. Some algorithms have few and others have many. Algorithms with fewer hyperparameters are easier to tune because there are less adjustments to consider.

Tuning machine learning algorithms is very difficult because it is often a nonintuitive, time-consuming, and systematic trial-and-error process. Difficulty is exacerbated because hyperparameters must be set manually before training can even begin. Tuning expertise can be enhanced by reading scholarly articles, industry books, online articles (e.g., Scikit-Learn documentation), watching YouTube videos, experience with data and data sets, diligence, hard work, and just plain practice.

© David Paper 2020
D. Paper, *Hands-on Scikit-Learn for Machine Learning Applications*,
https://doi.org/10.1007/978-1-4842-5373-1_5

Machine learning algorithms chosen for our tuning examples are not a coincidence. I chose them based on many hours of experimentation, reading, and insight. Algorithms that performed best for a given data set were included and those that performed poorly were not.

Scikit-Learn offers two vehicles for optimizing hyperparameter tuning: GridSearchCV and RandomizedSearchCV. *GridSearchCV* performs an exhaustive search over specified parameter values for an estimator (or machine learning algorithm) and returns the best performing hyperparametric combination.

So, all we need to do is specify the hyperparameters with which we want to experiment and their range of values, and GridSearchCV performs all possible combinations of hyperparameter values using cross-validation. As such, we naturally limit our choice of hyperparameters and their range of values. Theoretically, we can specify a set of parameter values for ALL hyperparameters of a model, but such a search consumes vast computer resources and time.

RandomizedSearchCV evaluates based on a predetermined subset of hyperparameters, randomly selects a chosen number of hyperparametric pairs from a given domain, and tests only those selected. RandomizedSearchCV tends to be less computationally expensive and time consuming because it doesn't evaluate every possible hyperparametric combination. This method greatly simplifies analysis without significantly sacrificing optimization. RandomizedSearchCV is often an excellent choice for high-dimensional data as it returns a good hyperparametric combination very quickly.

Tip   Tuning with GridSearchCV is suitable for an exhaustive search for the best performing hyperparameters given adequate computing resources. Tuning with RandomizedSearchCV is suitable for a good search or if tuning high-dimensional data.

Learning to tune classifiers can be accelerated by working through examples with a variety of data sets and classifiers. But, I also suggest following a structured process:

a) Always begin with default hyperparameters using baseline algorithms.

b) Experiment with training and test sizes.

c) Use dimensionality reduction when working with high-dimensional data.

d) Draw random samples when working with large data sets.

e) Scale data (where appropriate) to potentially increase performance.

f) Use GridSearchCV or RandomizedSearchCV to tune.

g) Once tuned with baseline algorithms, experiment with complex algorithms.

A *baseline algorithm* is one with few hyperparameters, so it is easy to tune. It also allows us to establish baseline performance on a predictive modeling problem. Using a baseline provides a point of comparison with more advanced algorithms that you evaluate later in the tuning process.

Tip   Begin tuning with a baseline algorithm (with its default hyperparameters) to establish baseline performance.

Once an optimally tuned algorithm is created on sample data, experiment with the full data set if adequate computer resources are available. Although such experiments can be costly, at least we have an excellent model to work with. Imagine the expense of trial-and-error experiments without sampling!

Since tuning is a complex endeavor, it is a good idea to learn by working with simple data sets. By simple, we mean *low-dimensional* and *small* data sets. First, tuning simple data allows experimentation without extensive computational expense or time. Trial-and-error tuning experimentation on simple data sets consumes relatively little computer and people time. Second, simple data sets are easy to work with and understand. Third, we can use many, if not all, hyperparameters, with GridSearchCV or RandomizedSearchCV.

Tuning Data Sets

We concentrate on four data sets: Iris, digits, banking, and wine. The Iris data set consists of 150 data elements representing three species of Iris. The digits data set consists of 1797 digit images. The banking data set consists of 41188 data elements representing client subscriptions. The wine data set consists of 178 data elements representing wine quality.

Tuning Iris Data

The code example shown in Listing 5-1 trains and tunes load_iris with
KNeighborsClassifier, GridSearchCV, and RandomizedSearchCV. Since we use the
humanfriendly package in the following code snippet and it is not usually preinstalled,
please install as follows:

```
pip install humanfriendly
```

Listing 5-1. Tuning Iris data with KNeighborsClassifier

```python
import numpy as np, humanfriendly as hf
import time
from sklearn.datasets import load_iris
from sklearn.model_selection import GridSearchCV
from sklearn.model_selection import RandomizedSearchCV
from sklearn.neighbors import KNeighborsClassifier
from sklearn.model_selection import train_test_split
from sklearn.metrics import accuracy_score
from sklearn.model_selection import cross_val_score

def get_scores(model, Xtrain, ytrain, Xtest, ytest):
    y_pred = model.predict(Xtrain)
    train = accuracy_score(ytrain, y_pred)
    y_pred = model.predict(Xtest)
    test = accuracy_score(ytest, y_pred)
    return train, test, model.__class__.__name__

def get_cross(model, data, target, groups=10):
    return cross_val_score(model, data, target, cv=groups)

def see_time(note):
    end = time.perf_counter()
    elapsed = end - start
    print (note, hf.format_timespan(elapsed, detailed=True))
```

```python
if __name__ == "__main__":
    br = '\n'
    iris = load_iris()
    X = iris.data
    y = iris.target
    targets = iris.target_names
    X_train, X_test, y_train, y_test = train_test_split(X, y, random_state=0)
    knn = KNeighborsClassifier()
    print (knn, br)
    distances = [1, 2, 3, 4, 5]
    k_range = list(range(1, 31))
    leaf = [10]
    param_grid = dict(n_neighbors=k_range, p=distances, leaf_size=leaf)
    start = time.perf_counter()
    grid = GridSearchCV(knn, param_grid, cv=10, scoring='accuracy')
    grid.fit(X, y)
    see_time('GridSearchCV total tuning time:')
    bp = grid.best_params_
    print ()
    print ('best parameters:')
    print (bp, br)
    knn_best = KNeighborsClassifier(**bp).fit(X_train, y_train)
    train, test, name = get_scores(knn_best, X_train, y_train, X_test, y_test)
    print (name, 'train/test scores (GridSearchCV):')
    print (train, test, br)
    scores = get_cross(knn, X, y)
    print ('cross-val scores:')
    print (scores, br)
    print ('avg cross-val score:', np.mean(scores), br)
    d = grid.cv_results_
    print ('mean grid score:', np.mean(d['mean_test_score']), br)
    vector = [[3, 5, 4, 2]]
    vectors = [[2, 5, 3, 5], [1, 4, 2, 1]]
    y_pred = knn_best.predict(vector)
    print (targets[y_pred])
```

```
    y_preds = knn_best.predict(vectors)
    print (targets[y_preds], br)
    start = time.perf_counter()
    rand = RandomizedSearchCV(knn, param_grid, cv=10, random_state=0,
                              scoring='accuracy', n_iter=10)
    rand.fit(X, y)
    see_time('RandomizedSearchCV total tuning time:')
    bp = rand.best_params_
    print()
    print ('best parameters:')
    print (bp, br)
    knn_best = KNeighborsClassifier(**bp).fit(X_train, y_train)
    train, test, name = get_scores(knn_best, X_train, y_train, X_test, y_test)
    print (name, 'train/test scores (RandomizedSearchCV):')
    print (train, test)
```

Go ahead and execute the code from Listing 5-1. Remember that you can find the example from the book's example download. You don't need to type the example by hand. It's easier to access the example download and copy/paste.

Your output from executing Listing 5-1 should resemble the following:

```
KNeighborsClassifier(algorithm='auto', leaf_size=30, metric='minkowski',
                     metric_params=None, n_jobs=None, n_neighbors=5, p=2,
                     weights='uniform')

GridSearchCV total tuning time: 7 seconds and 388.94 milliseconds

best parameters:
{'leaf_size': 10, 'n_neighbors': 6, 'p': 3}

KNeighborsClassifier train/test scores (GridSearchCV):
0.9732142857142857 0.9736842105263158

cross-val scores:
[1.         0.93333333 1.         1.         0.86666667 0.93333333
 0.93333333 1.         1.         1.         ]

avg cross-val score: 0.9666666666666668
```

mean grid score: 0.9673333333333333

['versicolor']
['versicolor' 'setosa']

RandomizedSearchCV total tuning time: 473.88 milliseconds

best parameters:
{'p': 3, 'n_neighbors': 13, 'leaf_size': 10}

KNeighborsClassifier train/test scores (RandomizedSearchCV):
0.9642857142857143 0.9736842105263158

The code begins by importing GridSearchCV, RandomizedSearchCV, cross_val_score, and other requisite packages. Function get_scores returns train-test accuracy and algorithm name. Function get_cross returns cross-validation score. Function see_time returns elapsed time. The main block loads the data and splits it into train-test subsets. The hyperparameters for KNeighborsClassifier are then displayed.

Tip   It is always a good idea to display an algorithm's hyperparameters before tuning.

We tune KNeighborsClassifier by adjusting *p, leaf_size,* and *n_neighbors.*

Tip   Scikit-Learn and other online documentation are good places to learn more about an algorithm's hyperparameters.

---

p is the power parameter for the Minkowski metric that adjusts distances. Adjust *leaf_size* to reduce number of candidates for the neighbors. Adjust *n_neighbors* to control number of neighbors.

The code continues by creating the parameter grid for GridSearchCV. We want to test distances (or p) from 1 to 5, n_neighbors from 1 to 31, and leaf_size of 10. By assigning the list [10] to leaf_size, we override its default value.

Tip   If a hyperparameter is not included in the search, its default value is used. If a single value (as a list) is included in the search, its value overrides the default but doesn't add computational expense to the search.

Next, we tune with GridSearchCV. Notice we use X and y because GridSearchCV conducts its own cross-validation of the data set. We continue by displaying the best parameters. We then use the best parameters with KNeighborsClassifier.

The results are excellent because the score is over 97% with an almost perfect fit. We run cross-validation on the train data to get an idea of how well an algorithm should perform. The cross-validation score is a bit under 97%, which means that our results are solid. Accuracy for a tuning experiment should approximate or exceed the cross-validation score.

Tip   The cross-validation score approximates the best performance we can attain from an algorithm. So, our performance should be close or hopefully better.

We then calculate and display average score from GridSearchCV. With our tuned model, we make some predictions. The code concludes by tuning using RandomizedSearchCV with the same parameter grid. Scores are almost identical, but elapsed time is much better with RandomizedSearchCV!

Tuning Digits Data

The code example shown in Listing 5-2 trains and tunes load_digits with KNeighborsClassifier, LogisticRegression, and GridSearchCV.

Listing 5-2. Tuning digits data with two algorithms

```
import numpy as np, humanfriendly as hf
import time
from sklearn.datasets import load_digits
from sklearn.model_selection import train_test_split,\
    cross_val_score
from sklearn.linear_model import LogisticRegression
from sklearn.neighbors import KNeighborsClassifier
from sklearn.metrics import accuracy_score
from sklearn.model_selection import GridSearchCV
```

```python
def get_scores(model, Xtrain, ytrain, Xtest, ytest):
    y_pred = model.predict(Xtrain)
    train = accuracy_score(ytrain, y_pred)
    y_pred = model.predict(Xtest)
    test = accuracy_score(ytest, y_pred)
    return train, test, model.__class__.__name__

def get_cross(model, data, target, groups=10):
    return cross_val_score(model, data, target, cv=groups)

def see_time(note):
    end = time.perf_counter()
    elapsed = end - start
    print (note, hf.format_timespan(elapsed, detailed=True))

if __name__ == "__main__":
    br = '\n'
    digits = load_digits()
    X = digits.data
    y = digits.target
    X_train, X_test, y_train, y_test = train_test_split(X, y, random_state=0)
    knn = KNeighborsClassifier().fit(X_train, y_train)
    print (knn, br)
    train, test, name = get_scores(knn, X_train, y_train, X_test, y_test)
    knn_name, acc1, acc2 = name, train, test
    print (str(knn_name) + ':')
    print ('train:', np.round(acc1, 2),
           'test:', np.round(acc2, 2), br)
    param_grid = {'n_neighbors': np.arange(1, 31, 2),
                  'metric': ['euclidean', 'cityblock']}
    start = time.perf_counter()
    grid = GridSearchCV(knn, param_grid, cv=5, n_jobs=-1)
    grid.fit(X, y)
    see_time('GridSearchCV total tuning time:')
    best_params = grid.best_params_
    print (best_params, br)
    knn_tuned = KNeighborsClassifier(**best_params)
```

```
knn_tuned.fit(X_train, y_train)
train, test, name = get_scores(knn_tuned, X_train, y_train, X_test, y_test)
knn_name, acc1, acc2 = name, train, test
print (knn_name + ' (tuned):')
print ('train:', np.round(acc1, 2),
       'test:', np.round(acc2, 2), br)
lr = LogisticRegression(random_state=0, max_iter=4000,
                        multi_class='auto', solver='lbfgs')
print (lr, br)
lr.fit(X_train, y_train)
train, test, name = get_scores(lr, X_train, y_train, X_test, y_test)
lr_name, acc1, acc2 = name, train, test
print (lr_name + ':')
print ('train:', np.round(acc1, 2),
       'test:', np.round(acc2, 2), br)
param_grid = {'penalty': ['l2'],
              'solver': ['newton-cg', 'lbfgs', 'sag'],
              'max_iter': [4000], 'multi_class': ['auto'],
              'C': [0.001, 0.01, 0.1]}
start = time.perf_counter()
grid = GridSearchCV(lr, param_grid, cv=5, n_jobs=-1)
grid.fit(X, y)
see_time('GridSearchCV total tuning time:')
bp = grid.best_params_
print (bp)
lr_tuned = LogisticRegression(**bp, random_state=0)
lr_tuned.fit(X_train, y_train)
train, test, name = get_scores(lr_tuned, X_train, y_train, X_test, y_test)
lr_name, acc1, acc2 = name, train, test
print (lr_name + ' (tuned):')
print ('train:', np.round(acc1, 2),
       'test:', np.round(acc2, 2), br)
print ('cross-validation score knn:')
knn = KNeighborsClassifier()
scores = get_cross(knn, X, y)
print (np.mean(scores))
```

Your output from executing Listing 5-2 should resemble the following:

```
KNeighborsClassifier(algorithm='auto', leaf_size=30,
        metric='minkowski', metric_params=None, n_jobs=None,
        n_neighbors=5, p=2, weights='uniform')

KNeighborsClassifier:
train: 0.99 test: 0.98

GridSearchCV total tuning time: 9 seconds and 609.98 milliseconds
{'metric': 'euclidean', 'n_neighbors': 3}

KNeighborsClassifier (tuned):
train: 0.99 test: 0.99

LogisticRegression(C=1.0, class_weight=None, dual=False,
        fit_intercept=True, intercept_scaling=1, max_iter=4000,
        multi_class='auto', n_jobs=None, penalty='l2',
        random_state=0, solver='lbfgs', tol=0.0001, verbose=0,
        warm_start=False)

LogisticRegression:
train: 1.0 test: 0.95

GridSearchCV total tuning time: 12 seconds and 708.79 milliseconds
{'C': 0.01, 'max_iter': 4000, 'multi_class': 'auto', 'penalty': 'l2',
'solver': 'lbfgs'}
LogisticRegression (tuned):
train: 0.99 test: 0.96

cross-validation score knn:
0.9739482872546906
```

The code begins by importing requisite packages. Function get_scores returns accuracy scores and model name. Function see_time returns elapsed time. The main block loads digit data into X and y and splits it into train-test subsets. Next, KNeighborsClassifier (with *default* hyperparameters) trains on the data, and results are displayed.

The code continues by tuning with GridSearchCV. For the tuning experiment, KNeighborsClassifier is the model, and we adjust *n_neighbors* and *metric* hyperparameters. From my experience and research, the number of neighbors is the most important hyperparameter to adjust for KNeighborsClassifier. Hyperparameter *metric* is the distance to use for the tree.

Tuning improved performance because we have an ideal fit. Of course, load_digits is heavily preprocessed, which makes it easy to tune. However, tuning is so complex that it is a good idea to work with simple data sets to learn the fundamentals.

The code continues by tuning with LogisticRegression. The algorithm (with its default parameters) is too complex as the results indicate overfitting. That is, the algorithm trains data perfectly, but test set accuracy is quite a bit lower.

Tip   When test accuracy is quite a bit lower than train accuracy, the training algorithm is too complex for the data set so overfitting occurs.

Tuning LogisticRegression with GridSearchCV reduces overfitting, but it doesn't perform as well as KNeighborsClassifier on the data. The hyperparameters adjusted for this tuning experiment include *penalty, solver, max_iter,* and *C. The penalty* involves the type of regularization. The *solver* specifies the algorithm to use during the optimization process. The *max_iter* hyperparameter indicates the maximum number of iterations taken for the solver to converge. Finally, *C* indicates the inverse of regularization strength. Smaller values for C specify stronger regularization.

The code ends by conducting cross-validation with KNeighborsClassifier. We conducted cross-validation with this algorithm because it was the best performer on the data. *Cross-validation* is a resampling procedure used to evaluate machine learning model performance. In this case, our performance is approaching 99%, which is better than the cross-validation score. So, we are confident that performance of our best model (KNeighborsClassifier) is optimal. If cross-validation is very different than our best modeling experiment, we may be doing something wrong or need to continue tuning.

Tip   Given adequate computing resources, cross-validation is a great technique to test the veracity of your algorithms.

Although LogisticRegression didn't perform as well as KNeighborsClassifier, tuning did improve performance. That is, test performance adjusted up while train performance adjusted down. When tuning adjusts train and test performance toward each other, we are making progress. However, if we are still not satisfied with performance, we should continue experimentation. But, at least we are moving in a positive direction.

Tip   As tuning experimentation adjusts train and test scores toward each other, we know our tuning experiment is making progress.

Tuning Bank Data

The first code example shown in Listing 5-3 tunes a random sample drawn from the bank data set with svm.SVC.

Listing 5-3. Tuning a bank data random sample with svm.SVC

```
import numpy as np, humanfriendly as hf, random
import time
from sklearn.model_selection import train_test_split,\
    RandomizedSearchCV, cross_val_score
from sklearn.preprocessing import StandardScaler
from sklearn.svm import SVC

def get_scores(model, xtrain, ytrain, xtest, ytest):
    ypred = model.predict(xtest)
    train = model.score(xtrain, ytrain)
    test = model.score(xtest, y_test)
    name = model.__class__.__name__
    return (name, train, test)

def get_cross(model, data, target, groups=10):
    return cross_val_score(model, data, target, cv=groups)

def prep_data(data, target):
    d = [data[i] for i, _ in enumerate(data)]
    t = [target[i] for i, _ in enumerate(target)]
    return list(zip(d, t))

def create_sample(d, n, replace='yes'):
    if replace == 'yes': s = random.sample(d, n)
```

```
    else: s = [random.choice(d)
                for i, _ in enumerate(d) if i < n]
    Xs = [row[0] for i, row in enumerate(s)]
    ys = [row[1] for i, row in enumerate(s)]
    return np.array(Xs), np.array(ys)

def see_time(note):
    end = time.perf_counter()
    elapsed = end - start
    print (note, hf.format_timespan(elapsed, detailed=True))

if __name__ == "__main__":
    br = '\n'
    X = np.load('data/X_bank.npy')
    y = np.load('data/y_bank.npy')
    sample_size = 4000
    data = prep_data(X, y)
    Xs, ys = create_sample(data, sample_size, replace='no')
    Xs = StandardScaler().fit_transform(Xs)
    X_train, X_test, y_train, y_test = train_test_split\
                                    (Xs, ys, random_state=0)
    svm = SVC(gamma='scale', random_state=0)
    print (svm, br)
    svm.fit(X_train, y_train)
    svm_scores = get_scores(svm, X_train, y_train, X_test, y_test)
    print (svm_scores[0] + ' (train, test):')
    print (svm_scores[1], svm_scores[2], br)
    Cs = [0.0001, 0.001]
    param_grid = {'C': Cs}
    start = time.perf_counter()
    rand = RandomizedSearchCV(svm, param_grid, cv=3, n_jobs = -1,
                            random_state=0, verbose=2, n_iter=2)
    rand.fit(X, y)
    see_time('RandomizedSearchCV total tuning time:')
    bp = rand.best_params_
    print (bp, br)
```

```
svm_tuned = SVC(**bp, gamma='scale', random_state=0)
svm_tuned.fit(X_train, y_train)
svm_scores = get_scores(svm_tuned, X_train, y_train, X_test, y_test)
print (svm_scores[0] + ' (train, test):')
print (svm_scores[1], svm_scores[2], br)
print ('cross-validation score:')
svm = SVC(gamma='scale')
scores = get_cross(svm, Xs, ys)
print (np.mean(scores))
```

Your output from executing Listing 5-3 should resemble the following:

```
SVC(C=1.0, cache_size=200, class_weight=None, coef0=0.0,
  decision_function_shape='ovr', degree=3, gamma='scale',
  kernel='rbf', max_iter=-1, probability=False, random_state=0,
  shrinking=True, tol=0.001, verbose=False)

SVC (train, test):
0.949 0.893

Fitting 3 folds for each of 2 candidates, totalling 6 fits
[Parallel(n_jobs=-1)]: Using backend LokyBackend with 8 concurrent workers.
[Parallel(n_jobs=-1)]: Done   3 out of   6 | elapsed:   55.4s
remaining:   55.4s
[Parallel(n_jobs=-1)]: Done   6 out of   6 | elapsed:   57.0s finished
RandomizedSearchCV total tuning time: 1 minute, 31 seconds and 171.06
milliseconds
{'C': 0.0001}

SVC (train, test):
0.891 0.875

cross-validation score:
0.9102441546509665
```

The code begins by importing RandomizedSearchCV, svm.SVC, and other requisite packages. Function get_scores returns accuracy scores and model name. Function prep_data prepares data for processing the sample in function create_sample. Function create_sample creates a random sample. Function see_time returns elapsed time.

The main block loads data, creates a sample of 4000 without replacement, and splits data into train-test subsets. Next, we scale data, train with svm.SVC (with default hyperparameters), and display results. The code continues by tuning svm.SVC with RandomizedSearchCV.

We only adjust the *C* hyperparameter, which is a regularization parameter that controls the trade-off between achieving a low error on training data and minimizing the norm of the weights. As we increase C, model complexity increases, which increases the chances of overfitting. Also notice that verbose is set to two (verbose=2).

The verbose parameter (*not* hyperparameter) controls the verbosity. The higher we set the number, the more messages we get. So, upon execution we notice messages about what is occurring during the tuning process. GridSearchCV also has a verbosity option.

By just tuning with two values of C, elapsed time is already over one minute! However, we seem to achieve a good fit. The cross-validation score confirms that we are doing well with svm.SVC, but could do even better with more tuning experimentation. That is, we might be able to squeeze a bit more performance out of svm.SVC with more experimentation.

Tip   If our best model test score is close to the cross-validation score, we don't need to continue tuning.

Through a lot of tuning experimentation, I was able to drastically reduce tuning complexity. I didn't begin with just two C values nor did I just tune with C. Moreover, I initially tried tuning with the entire data set but found it too computationally expensive, so I trained on a sample.

The next code example shown in Listing 5-4 tunes a random sample drawn from the bank data set with KNeighborsClassifier.

Listing 5-4. Tuning a bank data random sample with KNeighborsClassifier

```
import numpy as np, humanfriendly as hf, random
import time
from sklearn.model_selection import train_test_split,\
    RandomizedSearchCV, cross_val_score
from sklearn.neighbors import KNeighborsClassifier
```

```python
def get_scores(model, xtrain, ytrain, xtest, ytest):
    ypred = model.predict(xtest)
    train = model.score(xtrain, ytrain)
    test = model.score(xtest, y_test)
    name = model.__class__.__name__
    return (name, train, test)

def get_cross(model, data, target, groups=10):
    return cross_val_score(model, data, target, cv=groups)

def prep_data(data, target):
    d = [data[i] for i, _ in enumerate(data)]
    t = [target[i] for i, _ in enumerate(target)]
    return list(zip(d, t))

def create_sample(d, n, replace='yes'):
    if replace == 'yes': s = random.sample(d, n)
    else: s = [random.choice(d)
               for i, _ in enumerate(d) if i < n]
    Xs = [row[0] for i, row in enumerate(s)]
    ys = [row[1] for i, row in enumerate(s)]
    return np.array(Xs), np.array(ys)

def see_time(note):
    end = time.perf_counter()
    elapsed = end - start
    print (note, hf.format_timespan(elapsed, detailed=True))

if __name__ == "__main__":
    br = '\n'
    X = np.load('data/X_bank.npy')
    y = np.load('data/y_bank.npy')
    sample_size = 4000
    data = prep_data(X, y)
    Xs, ys = create_sample(data, sample_size, replace='no')
    X_train, X_test, y_train, y_test = train_test_split\
                                 (Xs, ys, random_state=0)
    knn = KNeighborsClassifier()
```

```
print (knn, br)
knn.fit(X_train, y_train)
knn_scores = get_scores(knn, X_train, y_train, X_test, y_test)
print (knn_scores[0] + ' (train, test):')
print (knn_scores[1], knn_scores[2], br)
param_grid = {'n_neighbors': np.arange(1, 31, 2),
              'metric': ['euclidean']}
start = time.perf_counter()
rand = RandomizedSearchCV(knn, param_grid, cv=3, n_jobs = -1,
                          random_state=0, verbose=2)
rand.fit(X, y)
see_time('RandomizedSearchCV total tuning time:')
bp = rand.best_params_
print (bp, br)
file = 'data/bp_bank'
np.save(file, bp)
knn_tuned = KNeighborsClassifier(**bp).fit(X_train, y_train)
knn_scores = get_scores(knn_tuned, X_train, y_train, X_test, y_test)
print (knn_scores[0] + ' (train, test):')
print (knn_scores[1], knn_scores[2], br)
print ('cross-validation score:')
knn = KNeighborsClassifier()
scores = get_cross(knn, Xs, ys)
print (np.mean(scores))
```

Your output from executing Listing 5-4 should resemble the following:

```
KNeighborsClassifier(algorithm='auto', leaf_size=30,
        metric='minkowski', metric_params=None, n_jobs=None,
        n_neighbors=5, p=2, weights='uniform')

KNeighborsClassifier (train, test):
0.927 0.906

Fitting 3 folds for each of 10 candidates, totalling 30 fits
[Parallel(n_jobs=-1)]: Using backend LokyBackend with 8 concurrent workers.
[Parallel(n_jobs=-1)]: Done  30 out of  30 | elapsed:   59.6s finished
```

```
RandomizedSearchCV total tuning time: 1 minute and 654.85 milliseconds
{'n_neighbors': 29, 'metric': 'euclidean'}

KNeighborsClassifier (train, test):
0.913 0.91

cross-validation score:
0.9032489046806542
```

The code begins by importing requisite packages. Function get_scores returns accuracy scores and model name. Function prep_data prepares data for processing the sample in function create_sample. Function create_sample creates a random sample. Function see_time returns elapsed time.

The main block loads data, creates a sample of 4000 without replacement, and splits data into train-test subsets. Next, we train with KNeighborsClassifier (with default hyperparameters) and display results. The code continues by tuning with RandomizedSearchCV. We adjust *n_neighbors* and force *metric* to *euclidean*. We also save best parameters for use in the next code example.

Through experimentation, I found that *euclidean* worked best. Tuning KNeighborsClassifier provided a much better fit than the model with default parameters. The cross-validation score confirms that we are doing well because it is very close to our test score. Notice that we did use 10 folds for cross-validation. My experience suggests that cross-validations of 5 or 10 seem to work well. However, be cautious as more cross-validations increase processing time.

The final code example in this section shown in Listing 5-5 models the *entire* bank data set using KNeighborsClassifier with best parameters garnered from the previous tuning exercise.

Listing 5-5. Tuning bank data with KNeighborsClassifier

```
import numpy as np, humanfriendly as hf, random
import time
from sklearn.model_selection import train_test_split
from sklearn.neighbors import KNeighborsClassifier

def get_scores(model, xtrain, ytrain, xtest, ytest):
    ypred = model.predict(xtest)
    train = model.score(xtrain, ytrain)
```

```python
        test = model.score(xtest, y_test)
        name = model.__class__.__name__
        return (name, train, test)

def see_time(note):
        end = time.perf_counter()
        elapsed = end - start
        print (note, hf.format_timespan(elapsed, detailed=True))

if __name__ == "__main__":
        br = '\n'
        X = np.load('data/X_bank.npy')
        y = np.load('data/y_bank.npy')
        bp = np.load('data/bp_bank.npy')
        bp = bp.tolist()
        print ('best parameters:')
        print (bp, br)
        X_train, X_test, y_train, y_test = train_test_split\
                                           (X, y, random_state=0)
        start = time.perf_counter()
        knn = KNeighborsClassifier(**bp)
        knn.fit(X_train, y_train)
        see_time('training time:')
        start = time.perf_counter()
        knn_scores = get_scores(knn, X_train, y_train, X_test, y_test)
        see_time('scoring time:')
        print ()
        print (knn_scores[0] + ' (train, test):')
        print (knn_scores[1], knn_scores[2])
```

Your output from executing Listing 5-5 should resemble the following:

```
best parameters:
{'n_neighbors': 29, 'metric': 'euclidean'}

training time: 461.58 milliseconds
scoring time: 10 seconds and 62.98 milliseconds
```

```
KNeighborsClassifier (train, test):
0.9154769997733968 0.9138584053607847
```

The code example imports requisite packages. Function get_scores returns accuracy scores and model name. Function see_time returns elapsed time. The main block loads bank data and best parameters for KNeighborsClassifier. Next, data is split into train-test subsets. The code ends by training the model using best parameters and displaying results.

Results indicate we achieved a really good fit and entire processing time is under eleven seconds! So, tuning with random samples is a great way to reduce computational expense.

Tip   Random sampling is a computationally inexpensive way to tune.

Tuning Wine Data

The first code example shown in Listing 5-6 leverages SGDClassifier and LinearDiscriminantAnalysis on load_wine.

Listing 5-6. Exploring wine data with two classifiers

```
import numpy as np, random
from sklearn.datasets import load_wine
from sklearn.preprocessing import StandardScaler
from sklearn.discriminant_analysis\
    import LinearDiscriminantAnalysis as LDA
from sklearn.linear_model import SGDClassifier
from sklearn.model_selection import train_test_split,\
    cross_val_score
from sklearn import metrics

def get_cross(model, data, target, groups=10):
    return cross_val_score(model, data, target, cv=groups)
```

```python
if __name__ == "__main__":
    br = '\n'
    wine = load_wine()
    X = wine.data
    y = wine.target
    X_train, X_test, y_train, y_test = train_test_split(X, y, random_state=0)
    lda = LDA().fit(X_train, y_train)
    print (lda, br)
    lda_name = lda.__class__.__name__
    y_pred = lda.predict(X_train)
    accuracy = metrics.accuracy_score(y_train, y_pred)
    accuracy = str(accuracy * 100) + '%'
    print (lda_name + ':')
    print ('train:', accuracy)
    y_pred_test = lda.predict(X_test)
    accuracy = metrics.accuracy_score(y_test, y_pred_test)
    accuracy = str(round(accuracy * 100, 2)) + '%'
    print ('test: ', accuracy, br)
    print ('cross-validation:')
    scores = get_cross(lda, X, y)
    print (np.mean(scores), br)
    n, ls = 100, []
    for i, row in enumerate(range(n)):
        rs = random.randint(1, 100)
        sgd = LDA().fit(X_train, y_train)
        y_pred = lda.predict(X_test)
        accuracy = metrics.accuracy_score(y_test, y_pred)
        ls.append(accuracy)
    avg = sum(ls) / len(ls)
    print ('MCS')
    print (avg, br)
    X = StandardScaler().fit_transform(X)
    X_train, X_test, y_train, y_test = train_test_split(X, y, random_state=0)
    sgd = SGDClassifier(max_iter=100, random_state=1)
    print (sgd, br)
```

```
    sgd.fit(X_train, y_train)
    sgd_name = sgd.__class__.__name__
    y_pred = sgd.predict(X_train)
    y_pred_test = sgd.predict(X_test)
    print (sgd_name + ':')
    print('train: {:.2%}'.format(metrics.accuracy_score\
                            (y_train, y_pred)))
    print('test:  {:.2%}\n'.format(metrics.accuracy_score\
                              (y_test, y_pred_test)))
    print ('cross-validation:')
    scores = get_cross(sgd, X, y)
    print (np.mean(scores), br)
    n, ls = 100, []
    for i, row in enumerate(range(n)):
        rs = random.randint(1, 100)
        sgd = SGDClassifier(max_iter=100).fit(X_train, y_train)
        y_pred = sgd.predict(X_test)
        accuracy = metrics.accuracy_score(y_test, y_pred)
        ls.append(accuracy)
    avg = sum(ls) / len(ls)
    print ('MCS:')
    print (avg)
```

Your output from executing Listing 5-6 should resemble the following:

```
LinearDiscriminantAnalysis(n_components=None, priors=None,
            shrinkage=None, solver='svd',
            store_covariance=False, tol=0.0001)

LinearDiscriminantAnalysis:
train: 100.0%
test:  97.78%

cross-validation:
0.9832989336085312

MCS
0.9777777777777754
```

```
SGDClassifier(alpha=0.0001, average=False, class_weight=None,
        early_stopping=False, epsilon=0.1, eta0=0.0,
        fit_intercept=True, l1_ratio=0.15,
        learning_rate='optimal', loss='hinge', max_iter=100,
        n_iter=None, n_iter_no_change=5, n_jobs=None,
        penalty='l2', power_t=0.5, random_state=1, shuffle=True,
        tol=None, validation_fraction=0.1, verbose=0,
        warm_start=False)
```

```
SGDClassifier:
train: 100.00%
test:  97.78%
```

```
cross-validation:
0.9616959064327485
```

```
MCS:
0.9966666666666663
```

The code begins by importing LinearDiscriminantAnalysis, SGDClassifier, and other requisite packages. The main block loads wine data, splits it into train-test sets, and trains with LinearDiscriminantAnalysis. The code continues by displaying accuracy scores, cross-validation, and MCS scores.

The cross-validation and MCS scores indicate that tuning most likely won't increase test accuracy for LinearDiscriminantAnalysis. So, we won't commence to tuning experiments. This example does, however, demonstrate that it is possible to obtain great accuracy scores without tuning experimentation. But, keep in mind that the load_wine data set is heavily processed. In industry, data is *rarely* this clean or as beautifully processed.

The next part of the code trains data with SGDClassifier. Notice that we scaled the data prior to splitting it into train-test subsets. I ran an experiment without scaling and obtained very poor results. So, SGDClassifier tends to benefit greatly from data scaling.

Scaling data for LinearDiscriminantAnalysis doesn't change results, so we didn't use scaled data for that experiment. Again, accuracy scores are excellent for SGDClassifier without tuning. Be cautious, however, when achieving perfect scores. I ran several experiments where I adjusted the random state parameter and the scores changed. Of course, the change wasn't very drastic, but train-test scores weren't perfect.

Although computationally expensive, MCS is a good indicator of how well an algorithm might perform on a data set. In the case of load_wine data, MCS is not a problem because the data set is small and isn't composed of high-dimensional data. Cross-validation is also an excellent indicator of algorithm performance, but it tends to be more conservative than MCS. It is also much less expensive computationally than MCS.

We can adjust the random state parameter to modify results. By changing random state on SGDClassifier from 0 to 1, test accuracy dips to 97.78%. This is another reason to run cross-validation and MCS (given adequate computational resources) to get an idea of a baseline accuracy for a given algorithm on a data set.

Tip   Adjusting the random state parameter changes scoring results, so it is always a good idea to run cross-validation to establish a stable baseline accuracy score.

The final code example shown in Listing 5-7 conducts an experiment on load_wine with a variety of classifiers. I included this example to demonstrate how one might explore the viability of algorithms for a given data set. Of course, we have to account for computational expense. But, given the resources for running such an experiment, we might save time and money in the long run.

Listing 5-7. Exploring wine data with a variety of classifiers

```
from sklearn.datasets import load_wine
from sklearn.neighbors import KNeighborsClassifier as knn
from sklearn.svm import SVC
from sklearn.gaussian_process import\
    GaussianProcessClassifier as gpc
from sklearn.gaussian_process.kernels import RBF as rbf
from sklearn.tree import DecisionTreeClassifier as dt
from sklearn.ensemble import RandomForestClassifier as rf,\
    AdaBoostClassifier as ada
from sklearn.naive_bayes import GaussianNB as gnb
from sklearn.discriminant_analysis import\
    QuadraticDiscriminantAnalysis as qda,\
    LinearDiscriminantAnalysis as lda
```

```
from sklearn.linear_model import SGDClassifier as sgd
from sklearn.preprocessing import StandardScaler
from sklearn.model_selection import train_test_split
from sklearn import metrics

if __name__ == "__main__":
    br = '\n'
    wine = load_wine()
    X = wine.data
    y = wine.target
    X = StandardScaler().fit_transform(X)
    X_train, X_test, y_train, y_test = train_test_split(
        X, y, test_size=.4, random_state=0)
    classifiers = [knn(3), qda(), lda(), gnb(),
                    SVC(kernel='linear', gamma='scale',
                        random_state=0),
                    ada(random_state=0), dt(random_state=0),
                    sgd(max_iter=100, random_state=0),
                    gpc(1.0 * rbf(1.0), random_state=0),
                    rf(random_state=0, n_estimators=100)]
    for clf in classifiers:
        clf.fit(X_train, y_train)
        train_score = clf.score(X_train, y_train)
        test_score = clf.score(X_test, y_test)
        name = clf.__class__.__name__
        print (name + '(train/test scores):')
        print (train_score, test_score)
```

Your output from executing Listing 5-7 should resemble the following:

```
KNeighborsClassifier(train/test scores):
0.9905660377358491 0.9027777777777778
QuadraticDiscriminantAnalysis(train/test scores):
0.9905660377358491 1.0
LinearDiscriminantAnalysis(train/test scores):
1.0 0.9722222222222222
```

```
GaussianNB(train/test scores):
0.9905660377358491 0.9444444444444444
SVC(train/test scores):
1.0 0.9722222222222222
AdaBoostClassifier(train/test scores):
1.0 0.9027777777777778
DecisionTreeClassifier(train/test scores):
1.0 0.9166666666666666
SGDClassifier(train/test scores):
1.0 0.9861111111111112
GaussianProcessClassifier(train/test scores):
1.0 0.9722222222222222
RandomForestClassifier(train/test scores):
1.0 0.9583333333333334
```

The code begins by loading requisite packages and a variety of classifiers including SGDClassifier and LinearDiscriminantAnalysis (demonstrated in the previous example). Keep in mind that this example is just an interesting experiment given appropriate computing resources.

The main block loads wine data, scales it, and splits it into train-test subsets. The code continues by creating a list of classifiers. The code concludes by iterating through the list of classifiers, training data with each classifier, and displaying accuracy scores.

From the results, the most viable classifiers for load_wine are SGDClassifier, LinearDiscriminantAnalysis, QuadraticDiscriminantAnalysis, svm.SVC, and GaussianProcessClassifier. RandomForestClassifier and GaussianNB also have potential, but less so than the ones listed first. QuadraticDiscriminantAnalysis produces an incredible fit with almost perfect scores!

We can tune the other algorithms to improve their performance, but my experience and this experiment tell me that we should work with the most promising algorithms to save time and money. This experiment is also a good way to get exposed to a variety of classification algorithms.

CHAPTER 6

Scikit-Learn Classifier Tuning from Complex Training Sets

Now that we have practiced tuning low-dimensional (or simple) data, we are ready to experiment tuning high-dimensional (or complex) data sets. *Low-dimensional* data consists of a limited number of features, whereas *high-dimensional* data consists of a very high number of features.

The term most commonly used to describe the dimensionality of a data set in machine learning literature is feature space. *Feature space* refers to the collection of features used to characterize the data set. That is, feature space refers to the n-dimensions where your variables live (not including a target variable if it is present).

Consistent with tuning low-dimensional data, we follow a structured process when tuning high-dimensional data:

a) Always begin with default hyperparameters using baseline algorithms.

b) Experiment with training and test sizes.

c) Use dimensionality reduction when working with high-dimensional data.

d) Draw random samples when working with large data sets.

e) Scale data (where appropriate) to potentially increase performance.

f) Use GridSearchCV or RandomizedSearchCV to tune.

g) Once tuned with baseline algorithms, experiment with complex algorithms.

© David Paper 2020
D. Paper, *Hands-on Scikit-Learn for Machine Learning Applications*,
https://doi.org/10.1007/978-1-4842-5373-1_6

Tuning Data Sets

We concentrate on three data sets: fetch_1fw_people, MNIST, and fetch_20newsgroups. The fetch_1fw_people data set contains 1288 face images and seven targets. Each face image is represented by a 50 × 37 matrix of pixels. The MNIST data set contains 70000 examples of handwritten digit images labeled from 0 to 9. Each digit is represented by a 28 × 28 matrix. The fetch_20newsgroups data set consists of approximately 18000 posts on 20 topics. Data is split into a training and testing sets. The split is based on messages posted before and after a specific date.

Tuning fetch_1fw_people

Face recognition is a *very* complex topic in machine learning. But, Scikit-Learn provides fetch_1fw_people that is a wonderful data set upon which to experiment and learn. Through experience and experimentation, I identified two Scikit-Learn algorithms – SGDClassifier and svm.SVC – that work relatively well with the data set.

The first code example shown in Listing 6-1 tunes data with SGDClassifier.

Listing 6-1. Tuning fetch_1fw_people with SGDClassifier

```
import numpy as np, humanfriendly as hf, warnings
import time
from sklearn.decomposition import PCA
from sklearn.model_selection import train_test_split,\
    GridSearchCV, cross_val_score
from sklearn.linear_model import SGDClassifier
from sklearn.metrics import classification_report

def see_time(note):
    end = time.perf_counter()
    elapsed = end - start
    print (note, hf.format_timespan(elapsed, detailed=True))

def get_cross(model, data, target, groups=10):
    return cross_val_score(model, data, target, cv=groups)
```

```python
if __name__ == "__main__":
    br = '\n'
    warnings.filterwarnings("ignore", category=DeprecationWarning)
    X = np.load('data/X_faces.npy')
    y = np.load('data/y_faces.npy')
    X_train, X_test, y_train, y_test = train_test_split(
        X, y, random_state=0)
    pca = PCA(n_components=0.95, whiten=True, random_state=1)
    pca.fit(X_train)
    X_train_pca = pca.transform(X_train)
    X_test_pca = pca.transform(X_test)
    pca_name = pca.__class__.__name__
    print ('<<' + pca_name + '>>')
    print ('features (before PCA):', X.shape[1])
    print ('features (after PCA):', pca.n_components_, br)
    sgd = SGDClassifier(max_iter=1000, tol=.001, random_state=0)
    sgd.fit(X_train_pca, y_train)
    y_pred = sgd.predict(X_test_pca)
    cr = classification_report(y_test, y_pred)
    print (cr)
    sgd_name = sgd.__class__.__name__
    param_grid = {'alpha': [1e-3, 1e-2, 1e-1, 1e0], 'max_iter': [1000],
                  'loss': ['log', 'perceptron'], 'penalty': ['l1'],
                  'tol': [.001]}
    grid = GridSearchCV(sgd, param_grid, cv=5)
    start = time.perf_counter()
    grid.fit(X_train_pca, y_train)
    see_time('training time:')
    print ()
    bp = grid.best_params_
    print ('best parameters:')
    print (bp, br)
    sgd = SGDClassifier(**bp, random_state=1)
```

```
sgd.fit(X_train_pca, y_train)
y_pred = sgd.predict(X_test_pca)
cr = classification_report(y_test, y_pred)
print (cr)
print ('cross-validation:')
scores = get_cross(sgd, X_train_pca, y_train)
print (np.mean(scores))
```

Go ahead and execute the code from Listing 6-1. Remember that you can find the example from the book's example download. You don't need to type the example by hand. It's easier to access the example download and copy/paste.

Your output from executing Listing 6-1 should resemble the following:

```
<<PCA>>
features (before PCA): 1850
features (after PCA): 135
```

	precision	recall	f1-score	support
0	0.89	0.57	0.70	28
1	0.80	0.78	0.79	63
2	0.83	0.62	0.71	24
3	0.73	0.89	0.80	132
4	0.55	0.55	0.55	20
5	0.88	0.32	0.47	22
6	0.67	0.73	0.70	33
micro avg	0.74	0.74	0.74	322
macro avg	0.76	0.64	0.67	322
weighted avg	0.76	0.74	0.73	322

```
training time: 7 seconds and 745.7 milliseconds

best parameters:
{'alpha': 0.001, 'loss': 'log', 'max_iter': 1000, 'penalty': 'l1', 'tol': 0.001}
```

	precision	recall	f1-score	support
0	0.91	0.71	0.80	28
1	0.79	0.79	0.79	63

2	0.71	0.71	0.71	24
3	0.84	0.86	0.85	132
4	0.48	0.75	0.59	20
5	0.83	0.45	0.59	22
6	0.72	0.79	0.75	33
micro avg	0.78	0.78	0.78	322
macro avg	0.76	0.72	0.73	322
weighted avg	0.79	0.78	0.78	322

```
cross-validation:
0.7808966616425951
```

The first code example begins by importing requisite packages. Function see_time returns elapsed time. The main block loads data into X and y, splits it into train-test subsets, and conducts PCA to reduce feature space dimensionality.

PCA is critical when tuning high-dimensional data because it *drastically* reduces computational expense with minimal information loss. The code then trains data with SGDClassifier (to obtain a baseline performance measure) and displays results. Next, tuning commences with GridSearchCV.

Tip   PCA is a critical tuning tool because it reduces dimensionality on high-dimensional data sets with minimal information loss, which results in drastically lower tuning time (or less computational expense).

We tune *alpha, max_iter, loss, penalty,* and *tol* hyperparameters. Hyperparameter *alpha* is the constant that multiplies the regularization term. Hyperparameter *max_iter* sets the maximum number of passes (or epochs) over training data. An *epoch* is one complete presentation of the data set to be learned by a machine.

Hyperparameter *loss* refers to the loss function used for the experiment. Machines learn by means of a *loss function,* which is a method for evaluating how well an algorithm models a given set of data. Hyperparameter *penalty* refers to the regularization term that is used by the model. Hyperparameter *tol* is the stopping criteria.

The two *most important hyperparameters* are alpha and penalty as they are directly related to the type and amount of regularization employed by the model.

The parameter grid is constructed next. Notice that alpha is the critical hyperparameter adjusted in this experiment. Through trial-and-error experiments, I determined that *l1* penalty was the best option, so I hard-coded it into the grid to reduce tuning time. Once tuned, SGDClassifier trains on the data with the best parameters and displays results. Finally, cross-validation is conducted to ensure that the model is performing at its best (which it is).

Tip   It is much easier (and faster) to conduct tuning experiments by varying one or two hyperparameters at a time and keeping the others constant by hard-coding their values.

The second code example shown in Listing 6-2 tunes with svm.SVC. From experience, I knew that svm.SVC outperformed SGDClassifier, but I wanted to demonstrate at least some of the rigor inherent in the experimental process of tuning by including the first code example in the chapter.

Listing 6-2. Tuning fetch_1fw_people with svm.SVC

```python
import numpy as np, humanfriendly as hf
import time
from sklearn.decomposition import PCA
from sklearn.model_selection import train_test_split,\
    GridSearchCV, cross_val_score
from sklearn.svm import SVC
from sklearn.metrics import classification_report
import matplotlib.pyplot as plt

def see_time(note):
    end = time.perf_counter()
    elapsed = end - start
    print (note, hf.format_timespan(elapsed, detailed=True))

def get_cross(model, data, target, groups=10):
    return cross_val_score(model, data, target, cv=groups)
```

```python
if __name__ == "__main__":
    br = '\n'
    X = np.load('data/X_faces.npy')
    y = np.load('data/y_faces.npy')
    images = np.load('data/faces_images.npy')
    targets = np.load('data/faces_targets.npy')
    _, h, w = images.shape
    n_images, n_features, n_classes = X.shape[0], X.shape[1],\
                                        len(targets)
    X_train, X_test, y_train, y_test = train_test_split(
        X, y, random_state=0)
    pca = PCA(n_components=0.95, whiten=True, random_state=0)
    pca.fit(X_train)
    components = pca.n_components_
    eigenfaces = pca.components_.reshape((components, h, w))
    X_train_pca = pca.transform(X_train)
    pca_name = pca.__class__.__name__
    print ('<<' + pca_name + '>>')
    print ('features (before PCA):', n_features)
    print ('features (after PCA):', components, br)
    X_i = np.array(eigenfaces[0].reshape(h, w))
    fig = plt.figure('eigenface')
    ax = fig.subplots()
    image = ax.imshow(X_i, cmap='bone')
    svm = SVC(random_state=0, gamma='scale')
    print (svm, br)
    svm.fit(X_train_pca, y_train)
    X_test_pca = pca.transform(X_test)
    y_pred = svm.predict(X_test_pca)
    cr = classification_report(y_test, y_pred)
    print (cr)
    svm_name = svm.__class__.__name__
    param_grid = {'C': [1e2, 1e3, 5e3], 'gamma': [0.001, 0.005, 0.01, 0.1],
                    'kernel': ['rbf'], 'class_weight': ['balanced']}
    grid = GridSearchCV(svm, param_grid, cv=5)
```

```
start = time.perf_counter()
grid.fit(X_train_pca, y_train)
see_time('training time:')
print ()
bp = grid.best_params_
print ('best parameters:')
print (bp, br)
svm = SVC(**bp)
svm.fit(X_train_pca, y_train)
y_pred = svm.predict(X_test_pca)
print ()
cr = classification_report(y_test, y_pred)
print (cr, br)
print ('cross-validation:')
scores = get_cross(svm, X_train_pca, y_train)
print (np.mean(scores), br)
file = 'data/bp_face'
np.save(file, bp)
bp = np.load('data/bp_face.npy')
bp = bp.tolist()
print ('best parameters:')
print (bp)
plt.show()
```

Your output from executing Listing 6-2 should resemble the following:

```
<<PCA>>
features (before PCA): 1850
features (after PCA): 135

SVC(C=1.0, cache_size=200, class_weight=None, coef0=0.0,
  decision_function_shape='ovr', degree=3, gamma='scale',
  kernel='rbf', max_iter=-1, probability=False, random_state=0,
  shrinking=True, tol=0.001, verbose=False)
```

	precision	recall	f1-score	support
0	1.00	0.43	0.60	28
1	0.83	0.87	0.85	63
2	0.94	0.62	0.75	24
3	0.71	0.97	0.82	132
4	1.00	0.70	0.82	20
5	1.00	0.36	0.53	22
6	0.96	0.73	0.83	33
micro avg	0.80	0.80	0.80	322
macro avg	0.92	0.67	0.74	322
weighted avg	0.84	0.80	0.78	322

training time: 18 seconds and 143.89 milliseconds

best parameters:
{'C': 100.0, 'class_weight': 'balanced', 'gamma': 0.005, 'kernel': 'rbf'}

	precision	recall	f1-score	support
0	1.00	0.64	0.78	28
1	0.76	0.92	0.83	63
2	0.91	0.88	0.89	24
3	0.88	0.92	0.90	132
4	0.74	0.85	0.79	20
5	1.00	0.64	0.78	22
6	0.90	0.85	0.88	33
micro avg	0.86	0.86	0.86	322
macro avg	0.89	0.81	0.84	322
weighted avg	0.87	0.86	0.86	322

cross-validation:
0.8393624737627647

best parameters:
{'C': 100.0, 'class_weight': 'balanced', 'gamma': 0.005, 'kernel': 'rbf'}

Listing 6-2 also displays Figure 6-1, which is the first eigenface created by PCA.

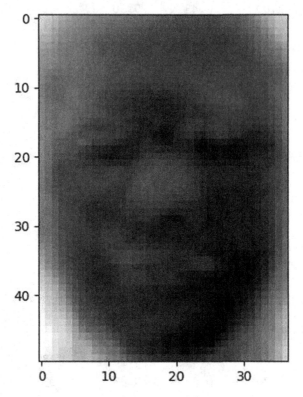

Figure 6-1. *First eigenface created by PCA*

The code begins by importing requisite packages. Function see_time returns elapsed time. The main block loads data into X and y, splits it into train-test subsets, and conducts PCA for dimensionality reduction. Baseline performance for svm.SVC is displayed for later comparison to the tuned svm.SVC score.

Tuning commences by constructing a grid with *C*, *gamma*, *kernel*, and *class_weight* hyperparameters. Hyperparameter *C* is the penalty parameter of the error term, so it is very important for tuning. Hyperparameter *gamma* is the kernel coefficient. Hyperparameter *kernel* specifies the kernel type to be used by the algorithm (e.g., linear). Hyperparameter *class_weight* is used to set the weight (or emphasis) of each class. Through experimentation, I found that the *rbf* kernel and *balanced* class weight were the best, so I hard-coded them into the grid.

My process of discovery is as follows: First, I kept all other hyperparameters constant and changed kernel to see the setting that yielded the best performance. Second, I kept kernel constant and changed class weight.

As you can tell by the grid, we vary C and gamma to improve performance. Once best parameters are determined, svm.SVC trains the data with them. Results are displayed along with cross-validation measures. We have done well with svm.SVC since we performed significantly better than the cross-validation score. We display the first eigenface from dimensionality reduction for completeness. Finally, best parameters are saved (and displayed).

Tuning MNIST

MNIST is not a large data set with 70000 examples, but it has a high-dimensional feature space consisting of 784 features. Such feature space complexity increases computational expense, so we must take this into account when running experiments with computationally expensive algorithms like svm.SVC.

The first code example in Listing 6-3 tunes MNIST with RandomForestClassifier and ExtraTreesClassifier. These algorithms have numerous hyperparameters, but we only adjust a few. I was able to greatly simplify tuning from my experience with these algorithms. You can experiment further, but computational expense increases greatly as you adjust additional hyperparameters.

Listing 6-3. Tuning with RandomForestClassifier and ExtraTreesClassifier

```
import numpy as np, humanfriendly as hf, random
import time
from sklearn.model_selection import train_test_split
from sklearn.model_selection import RandomizedSearchCV,\
    cross_val_score
from sklearn.ensemble import RandomForestClassifier,\
    ExtraTreesClassifier

def get_scores(model, xtrain, ytrain, xtest, ytest):
    ypred = model.predict(xtest)
    train = model.score(xtrain, ytrain)
    test = model.score(xtest, y_test)
```

```python
    name = model.__class__.__name__
    return (name, train, test)

def get_cross(model, data, target, groups=10):
    return cross_val_score(model, data, target, cv=groups)

def prep_data(data, target):
    d = [data[i] for i, _ in enumerate(data)]
    t = [target[i] for i, _ in enumerate(target)]
    return list(zip(d, t))

def create_sample(d, n, replace='yes'):
    if replace == 'yes': s = random.sample(d, n)
    else: s = [random.choice(d) for i, _ in enumerate(d) if i < n]
    Xs = [row[0] for i, row in enumerate(s)]
    ys = [row[1] for i, row in enumerate(s)]
    return np.array(Xs), np.array(ys)

def see_time(note):
    end = time.perf_counter()
    elapsed = end - start
    print (note, hf.format_timespan(elapsed, detailed=True))

if __name__ == "__main__":
    br = '\n'
    X_file = 'data/X_mnist'
    y_file = 'data/y_mnist'
    X = np.load('data/X_mnist.npy')
    y = np.load('data/y_mnist.npy')
    X = X.astype(np.float32)
    data = prep_data(X, y)
    sample_size = 7000
    Xs, ys = create_sample(data, sample_size)
    rf = RandomForestClassifier(random_state=0, n_estimators=100)
    print (rf, br)
    params = {'class_weight': ['balanced'], 'max_depth': [10, 30]}
    random = RandomizedSearchCV(rf, param_distributions = params,
                                cv=3, n_iter=2, random_state=0)
```

```
start = time.perf_counter()
random.fit(Xs, ys)
see_time('RandomizedSearchCV total tuning time:')
bp = random.best_params_
print (bp, br)
X_train, X_test, y_train, y_test = train_test_split(
    X, y, random_state=0)
rf = RandomForestClassifier(**bp, random_state=0, n_estimators=100)
start = time.perf_counter()
rf.fit(X_train, y_train)
rf_scores = get_scores(rf, X_train, y_train, X_test, y_test)
see_time('total time:')
print (rf_scores[0] + ' (train, test):')
print (rf_scores[1], rf_scores[2], br)
et = ExtraTreesClassifier(random_state=0, n_estimators=200)
print (et, br)
params = {'class_weight': ['balanced'], 'max_depth': [10, 30]}
random = RandomizedSearchCV(et, param_distributions = params,
                            cv=3, n_iter=2, random_state=0)
start = time.perf_counter()
random.fit(Xs, ys)
see_time('RandomizedSearchCV total tuning time:')
bp = random.best_params_
print (bp, br)
X_train, X_test, y_train, y_test = train_test_split(
    X, y, random_state=0)
et = ExtraTreesClassifier(**bp, random_state=0, n_estimators=200)
start = time.perf_counter()
et.fit(X_train, y_train)
et_scores = get_scores(et, X_train, y_train, X_test, y_test)
see_time('total time:')
print (et_scores[0] + ' (train, test):')
print (et_scores[1], et_scores[2], br)
print ('cross-validation (et):')
start = time.perf_counter()
```

```
scores = get_cross(rf, X, y)
see_time('total time:')
print (np.mean(scores), br)
file = 'data/bp_mnist_et'
np.save(file, bp)
bp = np.load('data/bp_mnist_et.npy')
bp = bp.tolist()
print ('best parameters:')
print (bp)
```

Your output from executing Listing 6-3 should resemble the following:

```
RandomForestClassifier(bootstrap=True, class_weight=None,
          criterion='gini', max_depth=None,
          max_features='auto', max_leaf_nodes=None,
          min_impurity_decrease=0.0, min_impurity_split=None,
          min_samples_leaf='deprecated', min_samples_split=2,
          min_weight_fraction_leaf='deprecated',
          n_estimators=100, n_jobs=None, oob_score=False,
          random_state=0, verbose=0, warm_start=False)

RandomizedSearchCV total tuning time: 13 seconds and 398.73 milliseconds
{'max_depth': 30, 'class_weight': 'balanced'}

total time: 32 seconds and 589.23 milliseconds
RandomForestClassifier (train, test):
0.9999809523809524 0.9701142857142857

ExtraTreesClassifier(bootstrap=False, class_weight=None,
          criterion='gini', max_depth=None, max_features='auto',
          max_leaf_nodes=None, min_impurity_decrease=0.0,
          min_impurity_split=None,
          min_samples_leaf='deprecated', min_samples_split=2,
          min_weight_fraction_leaf='deprecated',
          n_estimators=200, n_jobs=None, oob_score=False,
          random_state=0, verbose=0, warm_start=False)

RandomizedSearchCV total tuning time: 23 seconds and 342.93 milliseconds
{'max_depth': 30, 'class_weight': 'balanced'}
```

```
total time: 1 minute, 8 seconds and 270.59 milliseconds
ExtraTreesClassifier (train, test):
1.0 0.9732

cross-validation (et):
total time: 5 minutes, 40 seconds and 788.07 milliseconds
0.9692001937716965

best parameters:
{'max_depth': 30, 'class_weight': 'balanced'}
```

The code begins by importing requisite packages. Function get_scores returns accuracy scores and model name. Function get_cross returns cross-validation score. Function prep_data prepares data for function create_sample. Function create sample creates a random sample with or without replacement. Function see_time returns elapsed time. The main block loads data, creates a random sample, and instantiates algorithm RandomForestClassifier.

Tuning commences by constructing a grid with *class_weight* and *max_depth* hyperparameters. Hyperparameter *class_weight* is used to set the weight (or emphasis) of each class. Hyperparameter *max_depth* is used to establish the maximum depth of the tree. Through many hours of experimentation, I found that these two parameters were key to increasing performance. Tuning continues by leveraging RandomizedSearchCV to obtain the best parameters. Notice that tuning time is only a bit over thirteen seconds because the grid is very simple.

Now we can test RandomForestClassifier with best parameters. Notice that we include hyperparameter *n_estimators* in the algorithm along with best parameters. Hyperparameter *n_estimators* represents the number of trees in the forest and may be the most important hyperparameter for improving performance.

We include n_estimators in the algorithm (instead of putting it in the grid) for two reasons. First, it is such an important hyperparameter that we can save time by adjusting it outside a tuning experiment. That is, we can adjust it very easily without adding computational expense to the tuning experiment. However, increasing its value does add computational expense to processing the algorithm. Second, it must be included with this algorithm to avoid an annoying warning.

Tuning ExtraTreesClassifier follows the exact same logic with only one difference. We increase n_estimators to 200 trees. Notice that this increase causes processing time to more than double, but performance is better.

Finally, we run cross-validation (on ExtraTreesClassifier) and save the best parameters from ExtraTreesClassifier for future processing. From the cross-validation score, we know that our accuracy scores are solid. However, cross-validation consumes over 5 minutes of processing time! *You can comment out the cross-validation part of the code if you don't want to wait.*

On a positive note, cross-validation only needs to be executed once on an algorithm. I suggest that you run cross-validation before commencing a tuning experiment. You can then run trial-and-error experiments until you meet or exceed the cross-validation score.

Tip   Cross-validation need only be run once because it cannot be tuned.

Overall performance was good with accuracy over 97% with not too much overfitting. But, don't be lulled into a false sense of security by working through my tuning experiments. Tuning consumes a lot of time and patience. I can only give you examples and hints to help you become a more accomplished data scientist.

I highly recommend timing tuning experiments, especially ones that are computationally expensive (such as tuning with numerous hyperparameters over various ranges of values). Otherwise, it is very difficult to get a sense of how well your experiment is proceeding. When I first began tuning machine learning algorithms, I didn't time experiments. My progress was slow because I became very frustrated when I couldn't differentiate tuning experiments by elapsed time.

Tip   Always time tuning experiments to gauge progress.

The next code example shown in Listing 6-4 tunes MNIST with svm.SVC.

Listing 6-4. Tuning MNIST with svm.SVC

```
import numpy as np, humanfriendly as hf, random
import time
from sklearn.model_selection import train_test_split
from sklearn.model_selection import RandomizedSearchCV
from sklearn.decomposition import PCA
from sklearn.preprocessing import StandardScaler
from sklearn.svm import SVC
```

```python
def get_scores(model, xtrain, ytrain, xtest, ytest):
    ypred = model.predict(xtest)
    train = model.score(xtrain, ytrain)
    test = model.score(xtest, y_test)
    name = model.__class__.__name__
    return (name, train, test)

def prep_data(data, target):
    d = [data[i] for i, _ in enumerate(data)]
    t = [target[i] for i, _ in enumerate(target)]
    return list(zip(d, t))

def create_sample(d, n, replace='yes'):
    if replace == 'yes': s = random.sample(d, n)
    else: s = [random.choice(d) for i, _ in enumerate(d) if i < n]
    Xs = [row[0] for i, row in enumerate(s)]
    ys = [row[1] for i, row in enumerate(s)]
    return np.array(Xs), np.array(ys)

def see_time(note):
    end = time.perf_counter()
    elapsed = end - start
    print (note, hf.format_timespan(elapsed, detailed=True))

if __name__ == "__main__":
    br = '\n'
    X_file = 'data/X_mnist'
    y_file = 'data/y_mnist'
    X = np.load('data/X_mnist.npy')
    y = np.load('data/y_mnist.npy')
    X = X.astype(np.float32)
    data = prep_data(X, y)
    sample_size = 7000
    Xs, ys = create_sample(data, sample_size)
    pca = PCA(n_components=0.95, random_state=0)
    Xs = StandardScaler().fit_transform(Xs)
    Xs_reduced = pca.fit_transform(Xs)
```

```
X_train, X_test, y_train, y_test = train_test_split(
    Xs_reduced, ys, random_state=0)
svm = SVC(gamma='scale', random_state=0)
print (svm, br)
start = time.perf_counter()
svm.fit(X_train, y_train)
svm_scores = get_scores(svm, X_train, y_train, X_test, y_test)
print (svm_scores[0] + ' (train, test):')
print (svm_scores[1], svm_scores[2])
see_time('time:')
print ()
param_grid = {'C': [30, 35, 40], 'kernel': ['poly'],
              'gamma': ['scale'], 'degree': [3], 'coef0': [0.1]}
start = time.perf_counter()
rand = RandomizedSearchCV(svm, param_grid, cv=3, n_jobs = -1,
                          random_state=0, n_iter=3, verbose=2)
rand.fit(X_train, y_train)
see_time('RandomizedSearchCV total tuning time:')
bp = rand.best_params_
print (bp, br)
svm = SVC(**bp, random_state=0)
start = time.perf_counter()
svm.fit(X_train, y_train)
svm_scores = get_scores(svm, X_train, y_train, X_test, y_test)
print (svm_scores[0] + ' (train, test):')
print (svm_scores[1], svm_scores[2])
see_time('total time:')
```

Your output from executing Listing 6-4 should resemble the following:

```
SVC(C=1.0, cache_size=200, class_weight=None, coef0=0.0,
  decision_function_shape='ovr', degree=3, gamma='scale',
  kernel='rbf', max_iter=-1, probability=False, random_state=0,
  shrinking=True, tol=0.001, verbose=False)
```

```
SVC (train, test):
0.9845714285714285 0.9228571428571428
time: 13 seconds and 129.03 milliseconds

Fitting 3 folds for each of 3 candidates, totalling 9 fits
[Parallel(n_jobs=-1)]: Using backend LokyBackend with 8 concurrent workers.
[Parallel(n_jobs=-1)]: Done    4 out of    9 | elapsed:    14.0s
remaining:    17.6s
[Parallel(n_jobs=-1)]: Done    9 out of    9 | elapsed:    19.3s remaining:
0.0s
[Parallel(n_jobs=-1)]: Done    9 out of   9 | elapsed:    19.3s finished
RandomizedSearchCV total tuning time: 23 seconds and 824.72 milliseconds
{'kernel': 'poly', 'gamma': 'scale', 'degree': 3, 'coef0': 0.1, 'C': 30}

SVC (train, test):
1.0 0.9542857142857143
total time: 10 seconds and 810.06 milliseconds
```

Like the first MNIST tuning code example, we take a random sample. But, we also use PCA for dimensionality reduction because of the immense computational expense inherent with svm.SVC.

Tip   For computationally expensive algorithms, we recommend drawing a random sample *and* using PCA for dimensionality reduction to speed processing.

The code begins by importing requisite packages. We already talked about the functions in the last example, so we don't need to discuss it here.

The main block loads data and draws a random sample of 7000. PCA is used for dimensionality reduction with 5% information loss. Next, we scale training data because svm.SVC responds well to scaling. The code continues by splitting data into train-test subsets. Next, svm.SVC is trained with default parameters to gauge performance.

The code continues using RandomizedSearchCV to tune. We create a grid with hyperparameters *C*, *kernel*, *gamma*, *degree*, and *coef0*. We've already discussed hyperparameters C, kernel, and gamma, so we don't need to do it again here. Hyperparameter *degree* represents the degree of the polynomial kernel function. We include it because we chose *poly* for the kernel. Hyperparameter *coef0* is used in conjunction with *degree* for polynomial kernels.

Through experimentation, I found that hyperparameter C was the most important one to adjust. So, the grid only varies the values for C.

The code continues by using the best parameters from the tuning experiment with svm.SVC. We were able to increase test performance by quite a bit, but we still face overfitting.

We didn't include cross-validation for two reasons. First, svm.SVC didn't perform as well as ExtraTreeClassifier (so what's the point?). Second, it takes an extraordinary amount of time to run cross-validation on svm.SVC with MNIST.

Tuning fetch_20newsgroups

Like face recognition, text exploration is a *very* complex topic in machine learning. But, Scikit-Learn provides fetch_20newsgroups that is a wonderful data set upon which to experiment and learn.

Tuning complexity is greatly exacerbated because a pipelined model (with MultinomialNB and TfidfVectorizer) includes two sets of hyperparmeters (one from each algorithm).

Tuning MultinomialNB by itself is very easy because one need only adjust the *alpha* hyperparameter. Hyperparameter *alpha* allows us to adjust smoothing. However, tuning TfidfVectorizer is much more difficult as it includes numerous hyperparameters.

We encounter an even higher level of difficulty when tuning a pipelined model with RandomizedSearchCV because the names of the hyperparameters are different. Each hyperparameter from a pipelined model must be prefixed with the algorithm name so that RandomizedSearchCV can interpret correctly. This makes sense because algorithms can share the same hyperparameters.

The code example shown in Listing 6-5 tunes a pipelined model.

Listing 6-5. Tuning fetch_20newsgroups with a pipelined model

```
import numpy as np, humanfriendly as hf
import time
from sklearn.datasets import fetch_20newsgroups
from sklearn.feature_extraction.text import TfidfVectorizer
from sklearn.naive_bayes import MultinomialNB
from sklearn.pipeline import make_pipeline
from sklearn.metrics import f1_score
```

```python
from sklearn.model_selection import RandomizedSearchCV,\
    cross_val_score

def get_cross(model, data, target, groups=10):
    return cross_val_score(model, data, target, cv=groups)

def see_time(note):
    end = time.perf_counter()
    elapsed = end - start
    print (note, hf.format_timespan(elapsed, detailed=True))

if __name__ == "__main__":
    br = '\n'
    train = fetch_20newsgroups(subset='train')
    test = fetch_20newsgroups(subset='test')
    categories = ['rec.autos', 'rec.motorcycles', 'sci.space', 'sci.med']
    train = fetch_20newsgroups(subset='train', categories=categories,
                               remove=('headers', 'footers', 'quotes'))
    test = fetch_20newsgroups(subset='test', categories=categories,
                              remove=('headers', 'footers', 'quotes'))
    targets = train.target_names
    mnb = MultinomialNB()
    tf = TfidfVectorizer()
    print (mnb, br)
    print (tf, br)
    pipe = make_pipeline(tf, mnb)
    pipe.fit(train.data, train.target)
    labels = pipe.predict(test.data)
    f1 = f1_score(test.target, labels, average='micro')
    print ('f1_score', f1, br)
    print (pipe.get_params().keys(), br)
    param_grid = {'tfidfvectorizer__ngram_range': [(1, 1), (1, 2)],
                  'tfidfvectorizer__use_idf': [True, False],
                  'multinomialnb__alpha': [1e-2, 1e-3],
                  'multinomialnb__fit_prior': [True, False]}
    start = time.perf_counter()
```

```
rand = RandomizedSearchCV(pipe, param_grid, cv=3, n_jobs = -1,
                          random_state=0, n_iter=16, verbose=2)
rand.fit(train.data, train.target)
see_time('RandomizedSearchCV tuning time:')
bp = rand.best_params_
print ()
print ('best parameters:')
print (bp, br)
rbs = rand.best_score_
mnb = MultinomialNB(alpha=0.01)
tf = TfidfVectorizer(ngram_range=(1, 1), use_idf=False)
pipe = make_pipeline(tf, mnb)
pipe.fit(train.data, train.target)
labels = pipe.predict(test.data)
f1 = f1_score(test.target, labels, average='micro')
print ('f1_score', f1, br)
file = 'data/bp_news'
np.save(file, bp)
bp = np.load('data/bp_news.npy')
bp = bp.tolist()
print ('best parameters:')
print (bp, br)
start = time.perf_counter()
scores = get_cross(pipe, train.data, train.target)
see_time('cross-validation:')
print (np.mean(scores))
```

Your output from executing Listing 6-5 should resemble the following:

```
MultinomialNB(alpha=1.0, class_prior=None, fit_prior=True)

TfidfVectorizer(analyzer='word', binary=False,
        decode_error='strict', dtype=<class 'numpy.float64'>,
        encoding='utf-8', input='content', lowercase=True,
        max_df=1.0, max_features=None, min_df=1,
        ngram_range=(1, 1), norm='l2', preprocessor=None,
```

```
        smooth_idf=True, stop_words=None, strip_accents=None,
        sublinear_tf=False, token_pattern='(?u)\\b\\w\\w+\\b',
        tokenizer=None, use_idf=True, vocabulary=None)
```

f1_score 0.8440656565656567

dict_keys(['memory', 'steps', 'tfidfvectorizer', 'multinomialnb',
'tfidfvectorizer__analyzer', 'tfidfvectorizer__binary',
'tfidfvectorizer__decode_error', 'tfidfvectorizer__dtype',
'tfidfvectorizer__encoding', 'tfidfvectorizer__input',
'tfidfvectorizer__lowercase', 'tfidfvectorizer__max_df',
'tfidfvectorizer__max_features', 'tfidfvectorizer__min_df',
'tfidfvectorizer__ngram_range', 'tfidfvectorizer__norm',
'tfidfvectorizer__preprocessor', 'tfidfvectorizer__smooth_idf',
'tfidfvectorizer__stop_words', 'tfidfvectorizer__strip_accents',
'tfidfvectorizer__sublinear_tf', 'tfidfvectorizer__token_pattern',
'tfidfvectorizer__tokenizer', 'tfidfvectorizer__use_idf',
'tfidfvectorizer__vocabulary', 'multinomialnb__alpha',
'multinomialnb__class_prior', 'multinomialnb__fit_prior'])

Fitting 3 folds for each of 16 candidates, totalling 48 fits
[Parallel(n_jobs=-1)]: Using backend LokyBackend with 8 concurrent workers.
[Parallel(n_jobs=-1)]: Done 25 tasks | elapsed: 7.6s
[Parallel(n_jobs=-1)]: Done 48 out of 48 | elapsed: 12.4s finished
RandomizedSearchCV tuning time: 12 seconds and 747.04 milliseconds

best parameters:
{'tfidfvectorizer__use_idf': False, 'tfidfvectorizer__ngram_range': (1, 1),
'multinomialnb__fit_prior': False, 'multinomialnb__alpha': 0.01}

f1_score 0.8611111111111112

best parameters:
{'tfidfvectorizer__use_idf': False, 'tfidfvectorizer__ngram_range': (1, 1),
'multinomialnb__fit_prior': False, 'multinomialnb__alpha': 0.01}

cross-validation: 2 seconds and 750.36 milliseconds
0.8735201157292913
```
```

The code begins by importing requisite packages. Functions get_cross and see_time are next. The main block begins by creating train and test sets from the fetch_20newsgroups data set. Next, we create subcategories and split data into train-test subsets. The code continues by creating a baseline pipeline model and displaying f1_score for later comparison to the tuned model.

Possible hyperparameters of the pipelined model can be displayed with *pipe.get_params().keys()*. This is an important step because we must include the exact names for RandomizedSearchCV tuning.

Tip   You can (and should) display hyperparameters of a pipelined model with *model_name.get_params().keys()*.

The parameter grid is created with *tfidfvectorizer__ngram_range, tfidfvectorizer__use_idf, multinomialnb__alpha*, and *multinomialnb__fit_prior*.

Hyperparameter *multinomialnb__alpha* is exactly the same as alpha from MultinomialNB. The only difference is that prefix *multinomialnb* is included to inform RandomizedSearchCV the algorithm upon which it belongs. Hyperparameter *multinomialnb__fit_prior* indicates whether or not to learn class prior probabilities.

Hyperparameters *tfidfvectorizer__ngram_range* and *tfidfvectorizer__use_idf* belong to algorithm TfidfVectorizer as indicated by their prefixes. *ngram_range* indicates the upper and lower boundary of the range of n-values for different n-grams to be extracted from the document. *use_idf* enables or disables inverse-document-frequency reweighting.

Tuning commences with RandomizedSearchCV based on the parameter grid values. With tuning, we are able to increase performance to over 86%. However, cross-validation indicates that we can squeeze out a bit more performance from our model.

CHAPTER 7

Scikit-Learn Regression Tuning

Regression predictive modeling (or just *regression*) is the problem of learning the strength of association between independent variables (or features) and *continuous* dependent variables (or outcomes). Tuning regression algorithms is similar to tuning classification algorithms. That is, we adjust a model's hyperparameters until we arrive at an optimal solution.

The difference is that the goal of regression tuning is to reduce root mean squared error (RMSE), while the goal of classification tuning is to maximize accuracy. A benefit of *RMSE* is that units of the error score are the same as the predicted value. While regression predictions can be evaluated using RMSE, classification predictions cannot.

Tip The goal of regression tuning is to minimize RMSE.

Machine learning algorithms chosen for our tuning examples are not a coincidence. I chose them based on many hours of experimentation, reading, and insight. Algorithms that performed best for a given data set were included, and those that performed poorly were not.

For regression experiments in this chapter, we leverage GridSearchCV for tuning.

Tip Tuning with GridSearchCV is suitable for an exhaustive search for the best performing hyperparameters given adequate computing resources. Tuning with RandomizedSearchCV is suitable for a good search or if tuning high-dimensional data.

189

© David Paper 2020
D. Paper, *Hands-on Scikit-Learn for Machine Learning Applications*,
https://doi.org/10.1007/978-1-4842-5373-1_7

Learning to tune regression algorithms can be accelerated by working through examples with a variety of data sets and regressors. But, I also suggest following a structured process:

a) Always begin with default hyperparameters using baseline algorithms.

b) Experiment with training and test sizes.

c) Use dimensionality reduction when working with high-dimensional data.

d) Draw random samples when working with large data sets.

e) Scale data (where appropriate) to potentially increase performance.

f) Use GridSearchCV or RandomizedSearchCV to tune.

g) Once tuned with baseline algorithms, experiment with complex algorithms.

Tip   Begin tuning with a baseline algorithm (with its default hyperparameters) to establish baseline performance.

Tuning Data Sets

We concentrate on four data sets: tips, boston, and wine (red and white). tips data is composed of food server tips in restaurants and related factors including tip, price of meal, and time of day. boston data is composed of housing prices from various Boston locations. wine data is composed two data sets (red and white) that consist of variants of Portuguese Vinho Verde wine.

Tuning tips

The code example shown in Listing 7-1 calculates RMSE for a variety of regression algorithms based on unscaled and scaled data. Since tips is such a small data set, it is computationally inexpensive to run this type of experiment.

Listing 7-1. Calculating RMSE for tips data with regression algorithms

```
import numpy as np
from sklearn.model_selection import train_test_split
from sklearn.metrics import mean_squared_error
from sklearn.ensemble import RandomForestRegressor as rfr,\
    AdaBoostRegressor as ada, GradientBoostingRegressor as gbr
from sklearn.linear_model import LinearRegression as lr,\
    BayesianRidge as bay, Ridge as rr, Lasso as l,\
    LassoLars as ll, ElasticNet as en,\
    ARDRegression as ard, RidgeCV as rcv
from sklearn.svm import SVR
from sklearn.tree import DecisionTreeRegressor as dtr
from sklearn.neighbors import KNeighborsRegressor as knn
from sklearn.preprocessing import StandardScaler

def get_error(model, Xtest, ytest):
    y_pred = model.predict(Xtest)
    return np.sqrt(mean_squared_error(ytest, y_pred)),\
           model.__class__.__name__

if __name__ == "__main__":
    br = '\n'
    X = np.load('data/X_tips.npy')
    y = np.load('data/y_tips.npy')
    X_train, X_test, y_train, y_test = train_test_split(
        X, y, random_state=0)
    regressors = [lr(), bay(), rr(alpha=.5, random_state=0),
                  l(alpha=0.1, random_state=0), ll(), knn(),
                  ard(), rfr(random_state=0, n_estimators=100),
                  SVR(gamma='scale', kernel='rbf'),
                  rcv(fit_intercept=False), en(random_state=0),
                  dtr(random_state=0), ada(random_state=0),
                  gbr(random_state=0)]
    print ('unscaled:', br)
    for reg in regressors:
        reg.fit(X_train, y_train)
```

```
        rmse, name = get_error(reg, X_test, y_test)
        name = reg.__class__.__name__
        print (name + '(rmse):', end=' ')
        print (rmse)
    print ()
    print ('scaled:', br)
    scaler = StandardScaler()
    X_train_std = scaler.fit_transform(X_train)
    X_test_std = scaler.fit_transform(X_test)
    for reg in regressors:
        reg.fit(X_train_std, y_train)
        rmse, name = get_error(reg, X_test_std, y_test)
        name = reg.__class__.__name__
        print (name + '(rmse):', end=' ')
        print (rmse)
```

Go ahead and execute the code from Listing 7-1. Remember that you can find the example from the book's example download. You don't need to type the example by hand. It's easier to access the example download and copy/paste.

Your output from executing Listing 7-1 should resemble the following:

```
unscaled:

LinearRegression(rmse): 0.9474705746817206
BayesianRidge(rmse): 0.9245282337469829
Ridge(rmse): 0.9471900902779103
Lasso(rmse): 0.9158574785712037
LassoLars(rmse): 1.333812899498391
KNeighborsRegressor(rmse): 1.086204460049883
ARDRegression(rmse): 0.9264801346401996
RandomForestRegressor(rmse): 0.8850975551298138
SVR(rmse): 0.9441992099702836
RidgeCV(rmse): 0.9426372075893412
ElasticNet(rmse): 0.9307377813721578
DecisionTreeRegressor(rmse): 1.2994272932036561
AdaBoostRegressor(rmse): 0.932681302158466
GradientBoostingRegressor(rmse): 0.9112440690311495
```

scaled:

```
LinearRegression(rmse): 0.9007751177881488
BayesianRidge(rmse): 0.9096801291989541
Ridge(rmse): 0.9010890080377257
Lasso(rmse): 0.8785977911833892
LassoLars(rmse): 1.333812899498391
KNeighborsRegressor(rmse): 0.9613578099280607
ARDRegression(rmse): 0.8745960871430548
RandomForestRegressor(rmse): 0.893772251516372
SVR(rmse): 0.9749204385201592
RidgeCV(rmse): 3.1960055364135638
ElasticNet(rmse): 1.1310151423347359
DecisionTreeRegressor(rmse): 1.1835900827021861
AdaBoostRegressor(rmse): 0.986987944835978
GradientBoostingRegressor(rmse): 0.8908489427010696
```

The code begins by importing requisite packages and a variety of regression algorithms. Function get_error returns model name and RMSE. The main block begins by loading preprocessed tips data from NumPy files. Remember that we encoded tips data and saved it for future processing in Chapter 4.

Tip   Scikit-Learn allows you to experiment with a variety of algorithms to test performance without requiring contextual knowledge of them.

The code continues by splitting data into train-test subsets. Next, we create a list of regression algorithms. The code continues by training each algorithm on unscaled data and displaying results. The code then scales data, trains each algorithm on scaled data, and displays results.

Scaling data is a very important part of this experiment because many of the algorithms reported lower RMSE results than their unscaled brethren. The best performing algorithms with scaled data are Lasso and ARDRegression.

Tip   Scaling can be a very important technique during the tuning process.

So, the experiment was a success! It guided us to two algorithms upon which we can concentrate our tuning efforts.

The next code example shown in Listing 7-2 tunes tips with Lasso.

Listing 7-2. Tuning tips with Lasso

```python
import numpy as np, humanfriendly as hf
import time
from sklearn.preprocessing import StandardScaler
from sklearn.model_selection import train_test_split
from sklearn.linear_model import Lasso
from sklearn.model_selection import GridSearchCV,\
     cross_val_score
from sklearn.metrics import mean_squared_error

def get_error(model, Xtest, ytest):
    y_pred = model.predict(Xtest)
    return np.sqrt(mean_squared_error(ytest, y_pred)),\
           model.__class__.__name__

def see_time(note):
    end = time.perf_counter()
    elapsed = end - start
    print (note, hf.format_timespan(elapsed, detailed=True))

def get_cross(model, data, target, groups=10):
    return cross_val_score(model, data, target, cv=groups,
                           scoring='neg_mean_squared_error')

if __name__ == "__main__":
    br = '\n'
    X = np.load('data/X_tips.npy')
    y = np.load('data/y_tips.npy')
    X_train, X_test, y_train, y_test = train_test_split(
        X, y, random_state=0)
    scaler = StandardScaler()
    X_train_std = scaler.fit_transform(X_train)
    X_test_std = scaler.fit_transform(X_test)
```

```
lasso = Lasso(random_state=0, alpha=0.1)
print (lasso, br)
lasso.fit(X_train_std, y_train)
rmse, name = get_error(lasso, X_test_std, y_test)
print (name + '(rmse):', end=' ')
print (rmse, br)
alpha_lasso = [1e-1]
params = {'alpha': alpha_lasso, 'positive': [True, False],
          'max_iter': [10, 50, 100]}
grid = GridSearchCV(lasso, params, cv=5, n_jobs=-1, verbose=1)
start = time.perf_counter()
grid.fit(X_train, y_train)
see_time('training time:')
bp = grid.best_params_
print (bp, br)
lasso = Lasso(**bp, random_state=0).fit(X_train_std, y_train)
rmse, name = get_error(lasso, X_test_std, y_test)
print (name + '(rmse):', end=' ')
print (rmse, br)
start = time.perf_counter()
scores = get_cross(lasso, X, y)
see_time('cross-validation rmse:')
rmse = np.sqrt(np.mean(scores) * -1)
print (rmse)
```

Your output from executing Listing 7-2 should resemble the following:

```
Lasso(alpha=0.1, copy_X=True, fit_intercept=True, max_iter=1000,
   normalize=False, positive=False, precompute=False,
   random_state=0, selection='cyclic', tol=0.0001,
   warm_start=False)

Lasso(rmse): 0.8785977911833892
```

```
Fitting 5 folds for each of 6 candidates, totalling 30 fits
[Parallel(n_jobs=-1)]: Using backend LokyBackend with 8 concurrent workers.
[Parallel(n_jobs=-1)]: Done   30 out of   30 | elapsed:    2.1s finished
training time: 2 seconds and 246.86 milliseconds
{'alpha': 0.1, 'max_iter': 10, 'positive': True}

Lasso(rmse): 0.8781319871042923

cross-validation rmse: 8.58 milliseconds
1.0379804468729155
```

The code begins by importing requisite packages. Function get_error returns RMSE. Function see_time returns elapsed time. Function get_cross returns cross_validation RMSE.

The main block begins by loading preprocessed tips data. The code continues by splitting data into train-test subsets. Next, we scale data. We then train data with Lasso and display results for baseline comparison with the tuned RMSE.

Lasso is an algorithm that uses L1 penalty for regularization. We tune *alpha, positive,* and *max_iter* hyperparameters based on prior experimentation.

Hyperparameter *alpha* is the constant that multiplies the L1 penalty term. It is also the most important hyperparameter to tune with Lasso. Hyperparameter *positive* forces the coefficient to be positive. Hyperparameter *max_iter* represents the maximum number of iterations.

Tuning commences using GridSearchCV with grid *params*. With tuning, we were able to lower RMSE by a very small amount. Cross-validation reveals that we are doing very well.

Tip   Keep in mind that function get_error returns negative mean squared error, so we have to make the result positive by multiplying it by -1 and taking the square root of the result to get RMSE.

The next code example shown in Listing 7-3 tunes tips with ARDRegression.

Listing 7-3. Tuning tips with ARDRegression

```python
import numpy as np, humanfriendly as hf
import time
from sklearn.preprocessing import StandardScaler
from sklearn.model_selection import train_test_split
from sklearn.linear_model import ARDRegression
from sklearn.model_selection import GridSearchCV,\
    cross_val_score
from sklearn.metrics import mean_squared_error

def get_error(model, Xtest, ytest):
    y_pred = model.predict(Xtest)
    return np.sqrt(mean_squared_error(ytest, y_pred)),\
           model.__class__.__name__

def see_time(note):
    end = time.perf_counter()
    elapsed = end - start
    print (note, hf.format_timespan(elapsed, detailed=True))

def get_cross(model, data, target, groups=10):
    return cross_val_score(model, data, target, cv=groups,
                           scoring='neg_mean_squared_error')

if __name__ == "__main__":
    br = '\n'
    X = np.load('data/X_tips.npy')
    y = np.load('data/y_tips.npy')
    X_train, X_test, y_train, y_test = train_test_split(
        X, y, random_state=0)
    scaler = StandardScaler()
    X_train_std = scaler.fit_transform(X_train)
    X_test_std = scaler.fit_transform(X_test)
    ard = ARDRegression().fit(X_train_std, y_train)
    print (ard, br)
    rmse, name = get_error(ard, X_test_std, y_test)
```

```
print (name + '(rmse):', end=' ')
print (rmse, br)
iters = [50]
a1 = [1e5, 1e4]
a2 = [1e5, 1e4]
params = {'n_iter': iters, 'alpha_1': a1, 'alpha_2': a2}
grid = GridSearchCV(ard, params, cv=5, n_jobs=-1, verbose=1)
start = time.perf_counter()
grid.fit(X_train, y_train)
see_time('training time:')
bp = grid.best_params_
print (bp, br)
ard = ARDRegression(**bp).fit(X_train_std, y_train)
rmse, name = get_error(ard, X_test_std, y_test)
print (name + '(rmse):', end=' ')
print (rmse, br)
start = time.perf_counter()
scores = get_cross(ard, X, y)
see_time('cross-validation rmse:')
rmse = np.sqrt(np.mean(scores) * -1)
print (rmse)
```

Your output from executing Listing 7-3 should resemble the following:

```
ARDRegression(alpha_1=1e-06, alpha_2=1e-06, compute_score=False,
      copy_X=True, fit_intercept=True, lambda_1=1e-06,
      lambda_2=1e-06, n_iter=300, normalize=False,
      threshold_lambda=10000.0, tol=0.001, verbose=False)

ARDRegression(rmse): 0.8745960871430548

Fitting 5 folds for each of 4 candidates, totalling 20 fits
[Parallel(n_jobs=-1)]: Using backend LokyBackend with 8 concurrent workers.
[Parallel(n_jobs=-1)]: Done  20 out of  20 | elapsed:    3.5s finished
training time: 4 seconds and 286.03 milliseconds
{'alpha_1': 10000.0, 'alpha_2': 100000.0, 'n_iter': 50}
```

```
ARDRegression(rmse): 0.8645625277607758
```

```
cross-validation rmse: 4 seconds and 10.17 milliseconds
1.0376527153700184
```

The code begins by importing requisite packages. Function get_error returns RMSE. Function see_time returns elapsed time. Function get_cross returns cross_validation RMSE.

The main block begins by loading preprocessed tips data. The code continues by splitting data into train-test subsets. Next, we scale data. We then train data with ARDRegression and display results for baseline comparison with the tuned RMSE.

ARDRegression (Automatic Relevance Determination Regression) fits a regression model with Bayesian Ridge Regression. Estimation of the model is accomplished by iteratively maximizing the marginal log-likelihood of the observations.

We tune with *n_iter*, *alpha_1*, and *alpha_2*. Hyperparameter *n_iter* is the maximum number of iterations. Hyperparameter *alpha_1* is the shape parameter for the gamma distribution prior over the alpha parameter. Hyperparameter *alpha_2* is the inverse scale parameter (or rate parameter) for the gamma distribution prior over the alpha parameter.

We are able to reduce RMSE with tuning. Also, cross-validation reveals that we are doing very well.

Tuning boston

The code example shown in Listing 7-4 calculates RMSE for a variety of regression algorithms based on unscaled and scaled data. Since boston is a relatively small data set, it is computationally inexpensive to run this type of experiment.

Listing 7-4. Calculating RMSE for boston data with regression algorithms

```
import numpy as np
from sklearn.model_selection import train_test_split
from sklearn.metrics import mean_squared_error
from sklearn.ensemble import RandomForestRegressor as rfr,\
    AdaBoostRegressor as ada, GradientBoostingRegressor as gbr
from sklearn.linear_model import LinearRegression as lr,\
    BayesianRidge as bay, Ridge as rr, Lasso as l,\
    LassoLars as ll, ElasticNet as en,\
```

```
    ARDRegression as ard, RidgeCV as rcv
from sklearn.svm import SVR
from sklearn.tree import DecisionTreeRegressor as dtr
from sklearn.neighbors import KNeighborsRegressor as knn
from sklearn.preprocessing import StandardScaler

def get_error(model, Xtest, ytest):
    y_pred = model.predict(Xtest)
    return np.sqrt(mean_squared_error(ytest, y_pred)),\
           model.__class__.__name__

if __name__ == "__main__":
    br = '\n'
    X = np.load('data/X_boston.npy')
    y = np.load('data/y_boston.npy')
    X_train, X_test, y_train, y_test = train_test_split(
        X, y, random_state=0)
    regressors = [lr(), bay(), rr(alpha=.5, random_state=0),
                  l(alpha=0.1, random_state=0), ll(), knn(),
                  ard(), rfr(random_state=0, n_estimators=100),
                  SVR(gamma='scale', kernel='rbf'),
                  rcv(fit_intercept=False), en(random_state=0),
                  dtr(random_state=0), ada(random_state=0),
                  gbr(random_state=0)]
    print ('unscaled:', br)
    for reg in regressors:
        reg.fit(X_train, y_train)
        rmse, name = get_error(reg, X_test, y_test)
        name = reg.__class__.__name__
        print (name + '(rmse):', end=' ')
        print (rmse)
    print ()
    print ('scaled:', br)
    scaler = StandardScaler()
    X_train_std = scaler.fit_transform(X_train)
    X_test_std = scaler.fit_transform(X_test)
```

```
for reg in regressors:
    reg.fit(X_train_std, y_train)
    rmse, name = get_error(reg, X_test_std, y_test)
    name = reg.__class__.__name__
    print (name + '(rmse):', end=' ')
    print (rmse)
```

Your output from executing Listing 7-4 should resemble the following:

```
unscaled:

LinearRegression(rmse): 4.236710574387242
BayesianRidge(rmse): 4.317939916221959
Ridge(rmse): 4.243658717030716
Lasso(rmse): 4.300740333025026
LassoLars(rmse): 8.754893348840868
KNeighborsRegressor(rmse): 5.9934937623789
ARDRegression(rmse): 4.28415048500826
RandomForestRegressor(rmse): 3.37169151536684
SVR(rmse): 7.100029068343849
RidgeCV(rmse): 4.392246392993031
ElasticNet(rmse): 4.88844846745213
DecisionTreeRegressor(rmse): 4.346328232622458
AdaBoostRegressor(rmse): 3.652816906059683
GradientBoostingRegressor(rmse): 3.1941117128039194

scaled:

LinearRegression(rmse): 4.398269524691269
BayesianRidge(rmse): 4.419543929268475
Ridge(rmse): 4.400075160458176
Lasso(rmse): 4.489952156682322
LassoLars(rmse): 8.754893348840868
KNeighborsRegressor(rmse): 4.757936288305807
ARDRegression(rmse): 4.383622227159
RandomForestRegressor(rmse): 4.053037237125816
SVR(rmse): 5.083294658978756
RidgeCV(rmse): 22.34757636411328
ElasticNet(rmse): 5.277752330669967
```

```
DecisionTreeRegressor(rmse): 5.2796587719252726
AdaBoostRegressor(rmse): 4.100148076529094
GradientBoostingRegressor(rmse): 3.7490071027496015
```

The code begins by importing requisite packages and a variety of regression algorithms. Function get_error returns model name and RMSE. The main block begins by loading cleansed boston data from NumPy files. Remember that we cleansed boston data and saved it for future processing in Chapter 4.

The code continues by splitting data into train-test subsets. Next, we create a list of regression algorithms. The code continues by training each algorithm on unscaled data and displaying results. The code then scales data, trains each algorithm on scaled data, and displays results.

The best performing algorithms in this experiment are GradientBoostingRegressor and RandomForestRegressor (both with unscaled data). So, scaling data did not add value with this data set.

The next code example shown in Listing 7-5 tunes the boston data set with GradientBoostingRegressor.

Listing 7-5. Tuning boston data with GradientBoostingRegressor

```
import numpy as np, humanfriendly as hf, warnings, sys
import time
from sklearn.model_selection import train_test_split
from sklearn.ensemble import GradientBoostingRegressor
from sklearn.model_selection import GridSearchCV,\
    cross_val_score
from sklearn.metrics import mean_squared_error

def get_error(model, Xtest, ytest):
    y_pred = model.predict(Xtest)
    return np.sqrt(mean_squared_error(ytest, y_pred)),\
        model.__class__.__name__

def see_time(note):
    end = time.perf_counter()
    elapsed = end - start
    print (note, hf.format_timespan(elapsed, detailed=True))
```

```python
def get_cross(model, data, target, groups=10):
    return cross_val_score(model, data, target, cv=groups,
                           scoring='neg_mean_squared_error')

if __name__ == "__main__":
    br = '\n'
    if not sys.warnoptions:
        warnings.simplefilter('ignore')
    X = np.load('data/X_boston.npy')
    y = np.load('data/y_boston.npy')
    X_train, X_test, y_train, y_test = train_test_split(
        X, y, random_state=0)
    gbr = GradientBoostingRegressor(random_state=0)
    print (gbr, br)
    gbr.fit(X_train, y_train)
    rmse, name = get_error(gbr, X_test, y_test)
    print (name + '(rmse):', end=' ')
    print (rmse, br)
    loss = ['ls', 'lad', 'huber']
    lr = [1e-2, 1e-1, 1e-0]
    n_est = [150, 200, 300, 500]
    alpha = [0.9]
    params = {'loss': loss, 'learning_rate': lr,
              'n_estimators': n_est, 'alpha': alpha}
    grid = GridSearchCV(gbr, params, cv=5, n_jobs=-1,
                        verbose=1, refit=False)
    start = time.perf_counter()
    grid.fit(X_train, y_train)
    see_time('training time:')
    bp = grid.best_params_
    print (bp, br)
    gbr = GradientBoostingRegressor(**bp, random_state=0)
    gbr.fit(X_train, y_train)
    rmse, name = get_error(gbr, X_test, y_test)
    print (name + '(rmse):', end=' ')
    print (rmse, br)
```

```
start = time.perf_counter()
scores = get_cross(gbr, X, y)
see_time('cross-validation rmse:')
rmse = np.sqrt(np.mean(scores) * -1)
print (rmse)
```

Your output from executing Listing 7-5 should resemble the following:

```
GradientBoostingRegressor(alpha=0.9, criterion='friedman_mse',
            init=None, learning_rate=0.1, loss='ls',
            max_depth=3, max_features=None,
            max_leaf_nodes=None, min_impurity_decrease=0.0,
            min_impurity_split=None,
            min_samples_leaf='deprecated', min_samples_split=2,
            min_weight_fraction_leaf='deprecated',
            n_estimators=100, n_iter_no_change=None,
            presort='auto', random_state=0, subsample=1.0,
            tol=0.0001, validation_fraction=0.1, verbose=0,
            warm_start=False)

GradientBoostingRegressor(rmse): 3.1941117128039194

Fitting 5 folds for each of 36 candidates, totalling 180 fits
[Parallel(n_jobs=-1)]: Using backend LokyBackend with 8 concurrent workers.
[Parallel(n_jobs=-1)]: Done  34 tasks       | elapsed:    3.1s
[Parallel(n_jobs=-1)]: Done 180 out of 180 | elapsed:    9.1s finished
training time: 9 seconds and 170.11 milliseconds
{'alpha': 0.9, 'learning_rate': 0.1, 'loss': 'huber', 'n_estimators': 300}

GradientBoostingRegressor(rmse): 3.0839764165411934

cross-validation rmse: 3 seconds and 258.29 milliseconds
3.7929403445012064
```

The code begins by importing GradientBoostingRegressor as well as other requisite packages. *GradientBoostingRegressor* performs gradient boosting for regression by building an additive model in a forward-stage fashion.

Function get_error returns the RMSE and model name for a given algorithm. Function see_time returns elapsed time. Function get_cross returns the negative mean squared error.

The main block loads boston data, splits it into train-test subsets, and trains data with GradientBoostingRegressor. The code continues by displaying RMSE with default parameters to provide a baseline score for comparison to the tuned RMSE. Next, the model is tuned with hyperparameters *loss*, *learning_rate*, *n_estimators*, and *alpha*.

Hyperparameter *loss* is the loss function to be optimized. Hyperparameter *learning_rate* controls how much we adjust model learning with respect to the loss gradient. Hyperparameter *n_estimators* is the number of boosting stages to perform. Hypeparameter *alpha* is the alpha-quantile of the huber loss function.

Tuning enabled a reduction in RMSE. We end by running cross-validation. Since our tuned RMSE is lower than the cross-validation RMSE, we are in good shape.

Tip   You may have to occasionally reboot your computer as tuning requires an enormous amount of computing resources.

The final code example in this section (shown in Listing 7-6) tunes the boston data set with RandomForestRegressor.

Listing 7-6. Tuning boston data with RandomForestRegressor

```
import numpy as np, humanfriendly as hf, warnings, sys
import time
from sklearn.model_selection import train_test_split
from sklearn.ensemble import RandomForestRegressor
from sklearn.model_selection import GridSearchCV,\
     cross_val_score
from sklearn.metrics import mean_squared_error

def get_error(model, Xtest, ytest):
    y_pred = model.predict(Xtest)
    return np.sqrt(mean_squared_error(ytest, y_pred)),\
        model.__class__.__name__
```

```python
def see_time(note):
    end = time.perf_counter()
    elapsed = end - start
    print (note, hf.format_timespan(elapsed, detailed=True))

def get_cross(model, data, target, groups=10):
    return cross_val_score(model, data, target, cv=groups,
                           scoring='neg_mean_squared_error')

if __name__ == "__main__":
    br = '\n'
    if not sys.warnoptions:
        warnings.simplefilter('ignore')
    X = np.load('data/X_boston.npy')
    y = np.load('data/y_boston.npy')
    X_train, X_test, y_train, y_test = train_test_split(
        X, y, random_state=0)
    rfr = RandomForestRegressor(random_state=0)
    print (rfr, br)
    rfr.fit(X_train, y_train)
    rmse, name = get_error(rfr, X_test, y_test)
    print (name + '(rmse):', end=' ')
    print (rmse, br)
    n_est = [100, 500, 1000]
    boot = [True, False]
    params = {'n_estimators': n_est, 'bootstrap': boot}
    grid = GridSearchCV(rfr, params, cv=5, n_jobs=-1,
                        verbose=1, refit=False)
    start = time.perf_counter()
    grid.fit(X_train, y_train)
    see_time('training time:')
    bp = grid.best_params_
    print (bp, br)
    rfr = RandomForestRegressor(**bp, random_state=0)
    rfr.fit(X_train, y_train)
    rmse, name = get_error(rfr, X_test, y_test)
```

```
print (name + '(rmse):', end=' ')
print (rmse, br)
start = time.perf_counter()
scores = get_cross(rfr, X, y)
see_time('cross-validation rmse:')
rmse = np.sqrt(np.mean(scores) * -1)
print (rmse)
```

Your output from executing Listing 7-6 should resemble the following:

```
RandomForestRegressor(bootstrap=True, criterion='mse',
        max_depth=None, max_features='auto',
        max_leaf_nodes=None, min_impurity_decrease=0.0,
        min_impurity_split=None,
        min_samples_leaf='deprecated', min_samples_split=2,
        min_weight_fraction_leaf='deprecated',
        n_estimators='warn', n_jobs=None, oob_score=False,
        random_state=0, verbose=0, warm_start=False)

RandomForestRegressor(rmse): 3.5587794792757004

Fitting 5 folds for each of 6 candidates, totalling 30 fits
[Parallel(n_jobs=-1)]: Using backend LokyBackend with 8 concurrent workers.
[Parallel(n_jobs=-1)]: Done  30 out of  30 | elapsed:    8.3s finished
training time: 8 seconds and 453.84 milliseconds
{'bootstrap': True, 'n_estimators': 100}

RandomForestRegressor(rmse): 3.37169151536684

cross-validation rmse: 1 second and 845.76 milliseconds
3.6815463792891623
```

The code begins by importing RandomForestRegressor as well as other requisite packages. *RandomForestRegressor* fits a number of classifying decision trees on various subsamples of the data set and uses averaging to improve predictive accuracy and control overfitting.

Function get_error returns the RMSE and model name for a given algorithm. Function see_time returns elapsed time. Function get_cross returns the negative mean squared error.

The main block loads boston data, splits it into train-test subsets, and trains data with RandomForestRegressor. The code continues by displaying RMSE with default parameters to provide a baseline score for comparison to the tuned RMSE. Next, the model is tuned with hyperparameters *n_estimators* and *bootstrap*.

Hyperparameter *n_estimators* is the number of trees in the forest. Hyperparameter *bootstrap* determines whether bootstrap samples are used when building trees.

Tuning enabled a reduction in RMSE. We end by running cross-validation. Since our tuned RMSE is lower than the cross-validation RMSE, we are in good shape.

Tuning wine

By running an experiment similar to those shown in Listings 7-1 and 7-4, we found that RandomForestRegressor (with unscaled data) delivered the lowest RMSE for both red and white wine data. Go ahead and create your own experiments to verify our results if you wish.

The code example shown in Listing 7-7 tunes the red wine data set with RandomForestRegressor.

Listing 7-7. Tuning red wine data with RandomForestRegressor

```
import numpy as np, humanfriendly as hf
import time
from sklearn.model_selection import train_test_split
from sklearn.ensemble import RandomForestRegressor
from sklearn.model_selection import GridSearchCV,\
    cross_val_score
from sklearn.metrics import mean_squared_error

def get_error(model, Xtest, ytest):
    y_pred = model.predict(Xtest)
    return np.sqrt(mean_squared_error(ytest, y_pred)),\
        model.__class__.__name__

def see_time(note):
    end = time.perf_counter()
    elapsed = end - start
    print (note, hf.format_timespan(elapsed, detailed=True))
```

```python
def get_cross(model, data, target, groups=10):
    return cross_val_score(model, data, target, cv=groups,
                           scoring='neg_mean_squared_error')

if __name__ == "__main__":
    br = '\n'
    X = np.load('data/X_red.npy')
    y = np.load('data/y_red.npy')
    X_train, X_test, y_train, y_test = train_test_split(
        X, y, random_state=0)
    rfr = RandomForestRegressor(random_state=0, n_estimators=10)
    print (rfr, br)
    rfr.fit(X_train, y_train)
    rmse, name = get_error(rfr, X_test, y_test)
    print (name + '(rmse):', end=' ')
    print (rmse, br)
    n_est = [100, 500]
    boot = [True, False]
    params = {'n_estimators': n_est, 'bootstrap': boot}
    grid = GridSearchCV(rfr, params, cv=5, n_jobs=-1, verbose=1)
    start = time.perf_counter()
    grid.fit(X_train, y_train)
    see_time('training time:')
    bp = grid.best_params_
    print (bp, br)
    rfr = RandomForestRegressor(**bp, random_state=0)
    rfr.fit(X_train, y_train)
    rmse, name = get_error(rfr, X_test, y_test)
    print (name + '(rmse):', end=' ')
    print (rmse, br)
    start = time.perf_counter()
    scores = get_cross(rfr, X, y)
    see_time('cross-validation rmse:')
    rmse = np.sqrt(np.mean(scores) * -1)
    print (rmse)
```

Your output from executing Listing 7-7 should resemble the following:

```
RandomForestRegressor(bootstrap=True, criterion='mse',
                      max_depth=None, max_features='auto',
                      max_leaf_nodes=None,
                      min_impurity_decrease=0.0,
                      min_impurity_split=None,
                      min_samples_leaf='deprecated',
                      min_samples_split=2,
                      min_weight_fraction_leaf='deprecated',
                      n_estimators=10, n_jobs=None,
                      oob_score=False, random_state=0, verbose=0,
                      warm_start=False)
```

```
RandomForestRegressor(rmse): 0.626079068488957
```

```
Fitting 5 folds for each of 4 candidates, totalling 20 fits
[Parallel(n_jobs=-1)]: Using backend LokyBackend with 8 concurrent workers.
[Parallel(n_jobs=-1)]: Done  20 out of  20 | elapsed:    7.1s finished
training time: 7 seconds and 629.56 milliseconds
{'bootstrap': True, 'n_estimators': 100}
```

```
RandomForestRegressor(rmse): 0.5847897057917487
```

```
cross-validation rmse: 4 seconds and 804.96 milliseconds
0.6498982966515346
```

The code begins by importing requisite packages. Function get_error returns the RMSE and model name for a given algorithm. Function see_time returns elapsed time. Function get_cross returns the negative mean squared error.

The main block loads red wine data, splits it into train-test subsets, and trains data with RandomForestRegressor. The code continues by displaying RMSE with default parameters to provide a baseline score for comparison to the tuned RMSE. Next, the model is tuned with hyperparameters *n_estimators* and *bootstrap*.

Tuning enabled a reduction in RMSE. We end by running cross-validation. Since our tuned RMSE is lower than the cross-validation RMSE, we are in good shape.

The final code example shown in Listing 7-8 tunes the white wine data set with RandomForestRegressor.

Listing 7-8. Tuning white wine data with RandomForestRegressor

```python
import numpy as np, humanfriendly as hf
import time
from sklearn.model_selection import train_test_split
from sklearn.ensemble import RandomForestRegressor
from sklearn.model_selection import GridSearchCV,\
    cross_val_score
from sklearn.metrics import mean_squared_error

def get_error(model, Xtest, ytest):
    y_pred = model.predict(Xtest)
    return np.sqrt(mean_squared_error(ytest, y_pred)),\
           model.__class__.__name__

def see_time(note):
    end = time.perf_counter()
    elapsed = end - start
    print (note, hf.format_timespan(elapsed, detailed=True))

def get_cross(model, data, target, groups=10):
    return cross_val_score(model, data, target, cv=groups,
                           scoring='neg_mean_squared_error')

if __name__ == "__main__":
    br = '\n'
    X = np.load('data/X_white.npy')
    y = np.load('data/y_white.npy')
    X_train, X_test, y_train, y_test = train_test_split(
        X, y, random_state=0)
    rfr = RandomForestRegressor(random_state=0, n_estimators=10)
    print (rfr, br)
    rfr.fit(X_train, y_train)
    rmse, name = get_error(rfr, X_test, y_test)
    print (name + '(rmse):', end=' ')
    print (rmse, br)
    n_est = [100, 500]
    boot = [True, False]
```

211

```
params = {'n_estimators': n_est, 'bootstrap': boot}
grid = GridSearchCV(rfr, params, cv=5, n_jobs=-1, verbose=1)
start = time.perf_counter()
grid.fit(X_train, y_train)
see_time('training time:')
bp = grid.best_params_
print (bp, br)
rfr = RandomForestRegressor(**bp, random_state=0)
rfr.fit(X_train, y_train)
rmse, name = get_error(rfr, X_test, y_test)
print (name + '(rmse):', end=' ')
print (rmse, br)
start = time.perf_counter()
scores = get_cross(rfr, X, y)
see_time('cross-validation rmse:')
rmse = np.sqrt(np.mean(scores) * -1)
print (rmse)
```

Your output from executing Listing 7-8 should resemble the following:

```
RandomForestRegressor(bootstrap=True, criterion='mse',
          max_depth=None, max_features='auto',
          max_leaf_nodes=None, min_impurity_decrease=0.0,
          min_impurity_split=None,
          min_samples_leaf='deprecated', min_samples_split=2,
          min_weight_fraction_leaf='deprecated',n_estimators=10,
          n_jobs=None, oob_score=False, random_state=0,
          verbose=0, warm_start=False)

RandomForestRegressor(rmse): 0.6966098665124181

Fitting 5 folds for each of 4 candidates, totalling 20 fits
[Parallel(n_jobs=-1)]: Using backend LokyBackend with 8 concurrent workers.
[Parallel(n_jobs=-1)]: Done  20 out of  20 | elapsed:   18.7s finished
training time: 25 seconds and 709.64 milliseconds
{'bootstrap': True, 'n_estimators': 500}
```

```
RandomForestRegressor(rmse): 0.6728175517621279
```

```
cross-validation rmse: 1 minute, 24 seconds and 70.99 milliseconds
0.7183073387927801
```

The code begins by importing requisite packages. Function get_error returns the RMSE and model name for a given algorithm. Function see_time returns elapsed time. Function get_cross returns the negative mean squared error.

The main block loads white wine data, splits it into train-test subsets, and trains data with RandomForestRegressor. The code continues by displaying RMSE with default parameters to provide a baseline score for comparison to the tuned RMSE. Next, the model is tuned with hyperparameters *n_estimators* and *bootstrap*.

Tuning enabled a reduction in RMSE. We end by running cross-validation. Since our tuned RMSE is lower than the cross-validation RMSE, we are in good shape.

CHAPTER 8

Putting It All Together

The Journey

During our journey with machine learning, we worked with eleven data sets. Seven are classification data sets, while four are predictive modeling regression data sets.

Four of the classification data sets are simple. Three are complex. Simple data sets are those with few features. A data set with few features is typically referred to as one with a low-dimensional feature space. Complex data sets are those with many features. Such a data set is typically referred to as one with a high-dimensional feature space. All four of the predictive modeling regression (or just regression) data sets are simple in that their respective feature spaces have few features.

Feature space dimensionality in machine learning rests upon the notion that each feature represents one dimension. So, a feature set with a few features has low dimensionality and one with a lot of dimensions has high dimensionality. Feature space dimensionality is important in machine learning because a data set with a high-dimensional feature space is computationally expensive. That is, algorithmic machine learning with such data requires abundant computer resources.

Classification predicts the category the data belongs to, while regression predicts a numerical value based on previous observed data. So, classification is used to predict discrete responses like gender or a type of fruit. Regression is used to predict continuous values like housing prices or profits.

We continued our journey by demonstrating how machine learning algorithms learn from data. We began by training both simple and complex classification data sets with a variety of classification algorithms in Chapters 2 and 3. We then trained regression data sets with a variety of regression algorithms in Chapter 4.

For both classification and regression learning, we demonstrated how to make predictions from trained algorithms. Predictions allow us to see results from algorithmic training. With classification, we predict discreet targets based on new data.

© David Paper 2020
D. Paper, *Hands-on Scikit-Learn for Machine Learning Applications*,
https://doi.org/10.1007/978-1-4842-5373-1_8

With regression, we predict continuous outcomes based on new data. In both cases, we used test data split from our full data set as new data. However, predictions are only as good as our training.

To assess training performance, we must know how to measure learning. For classification, we showed you how to derive accuracy. Accuracy is the percentage of predictions that we got right. For regression, we showed you how to derive RMSE (root mean squared error). RMSE is the difference between predicted values and the observed ones. Simply, RMSE measures error.

Although we showed you how to train data with machine learning algorithms, it is possible to increase performance with model tuning. So, we showed you how to tune classification algorithms in Chapters 5 and 6 as well as regression algorithms in Chapter 7. As you know, tuning is a very complex, arduous, time-consuming, and experimental process. So, you need patience and fortitude to improve performance through tuning.

Tuning is also a very effective way to reduce overfitting. Overfitting is when an algorithm is memorizing data instead of learning from it. You know your model is overfitting when your training accuracy is a lot higher than your test accuracy.

The machine learning journey, however, is just beginning.

Value and Cost

Learning the technical side of machine learning is not enough. Data sets in industry tend to be large to extremely large. Even a large data set with a low-dimensional feature space can be computationally expensive. Imagine the computational expense of training an extremely large data set with a high-dimensional feature space! If tuning is included, computational expense and data scientist time can be exorbitant.

Data scientists want data in its natural raw state. They want to collect it from the source as it is generated. They don't want to access data from relational databases or data stored in various forms in legacy systems. However, current legacy systems more often than not store data in relational databases. Furthermore, new data is often placed into the same systems. Finally, organizations have rules in place concerning who accesses data, how much can be accessed, and when it can be accessed.

Data scientists want raw data as it's generated to match what actually happens in the natural world. That is, they want to mimic reality. Since the idea of machine learning is to *learn* from data, how, where, when, and why data is collected is paramount.

If the organization collects and processes data into its systems, the natural meaning (and timing) of the data is lost. And, data scientists may have great difficulty even getting the data they need.

So, not only do data scientists struggle to get data in its natural state, they must navigate all of the rules, security policies, and politics to access current or new data. Current organizational structures are not built to enable data scientists to get the data they want, when they want it, and in the form they want it. In addition, data scientists are a relatively new phenomenon in organizations. So, they tend to have less political clout, their role can be misunderstood, and what they do with data doesn't match what has been done in the past.

Less political clout makes it very difficult to convince those with the financial resources (or money people) to allocate more funding for expensive computing resources above and beyond current allocations. If money people misunderstand the role of data scientists and how resources were allocated in the past, data scientists may not get the resources they require. Moreover, data scientists get paid very well and are extremely well-educated. Naturally, someone that gets paid more tends to increase jealously and turf battles. In addition, data scientists can be viewed as know-it-alls because of their education, newness to the organization, and differing views of data.

So, what can be done to convince money people to budget for data scientists' data needs? First, we have to understand what is important to them. Second, we have to realize that they naturally want us to succeed because of the incredible rush to embrace machine learning in industry.

Only two things are critically important to money people – value and cost. *Value* is what determines the health and well-being of the organization. *Cost* is anything that detracts from the health and well-being of the organization. So, we must be able to present our case to money people that what we do is both valuable and reduces costs. We must also be cognizant that money people are not data scientists and most likely are not technically oriented.

The remainder of the chapter demonstrates value and cost to money people by presenting three complex code examples built in this book in very simple terms. Each code example is presented with output and explained simply with an emphasis on value and cost savings. The code is not explained. It is just included to demonstrate the complexity of algorithmic learning by showing what data scientists actually do. Finally, we chose the most complex data sets to prove that they can be explained to nontechnical people.

MNIST Value and Cost

We begin with the MNIST code example that was previously tuned, which is shown in Listing 8-1. We load the data, run the machine learning algorithm, and display results. We then explain results in business value and costs terms.

Listing 8-1. MNIST value and cost

```
import numpy as np, humanfriendly as hf
import time
from sklearn.model_selection import train_test_split
from sklearn.ensemble import ExtraTreesClassifier
from sklearn.metrics import confusion_matrix
import matplotlib.pyplot as plt
import seaborn as sns

def get_scores(model, xtrain, ytrain, xtest, ytest):
    ypred = model.predict(xtest)
    train = model.score(xtrain, ytrain)
    test = model.score(xtest, y_test)
    name = model.__class__.__name__
    return (name, train, test, ypred)

def see_time(note):
    end = time.perf_counter()
    elapsed = end - start
    print (note, hf.format_timespan(elapsed, detailed=True))

def find_misses(test, pred):
    return [i for i, row in enumerate(test) if row != pred[i]]

if __name__ == "__main__":
    br = '\n'
    X_file = 'data/X_mnist'
    y_file = 'data/y_mnist'
    X = np.load('data/X_mnist.npy')
    y = np.load('data/y_mnist.npy')
    X = X.astype(np.float32)
```

```
bp = np.load('data/bp_mnist_et.npy')
bp = bp.tolist()
X_train, X_test, y_train, y_test = train_test_split(
    X, y, random_state=0)
et = ExtraTreesClassifier(**bp, random_state=0, n_estimators=200)
start = time.perf_counter()
et.fit(X_train, y_train)
et_scores = get_scores(et, X_train, y_train, X_test, y_test)
see_time('total time:')
print (et_scores[0] + ' (train, test):')
print (et_scores[1], et_scores[2], br)
y_pred = et_scores[3]
cm = confusion_matrix(y_test, y_pred)
plt.figure(1)
ax = plt.axes()
sns.heatmap(cm.T, annot=True, fmt="d", cmap='gist_ncar_r', ax=ax)
ax.set_title(et_scores[0] + 'confustion matrix')
plt.xlabel('true value')
plt.ylabel('predicted value')
indx = find_misses(y_test, y_pred)
print ('pred', 'actual')
misses = [(y_pred[row], y_test[row], i)
          for i, row in enumerate(indx)]
[print (row[0], '  ', row[1]) for i, row in enumerate(misses)
 if i < 10]
print()
img_act = y_test[indx[0]]
img_pred = y_pred[indx[0]]
print ('actual', img_act)
print ('pred', img_pred)
text = str(img_pred)
test_images = X_test.reshape(-1, 28, 28)
plt.figure(2)
plt.imshow(test_images[indx[0]], cmap='gray', interpolation='gaussian')
plt.text(0, 0.05, text, color='r', bbox=dict(facecolor='white'))
```

```
title = str(img_act) + ' misclassified as ' + text
plt.title(title)
plt.show()
```

Your output from executing Listing 8-1 should resemble the following:

```
total time: 1 minute, 8 seconds and 650.43 milliseconds
ExtraTreesClassifier (train, test):
1.0 0.9732

pred actual
3.0     9.0
7.0     3.0
4.0     9.0
2.0     3.0
3.0     2.0
6.0     5.0
9.0     7.0
9.0     3.0
8.0     6.0
9.0     4.0

actual 9.0
pred 3.0
```

Listing 8-1 also displays Figures 8-1 and 8-2. Figure 8-1 shows the confusion matrix, and Figure 8-2 shows the first misclassification.

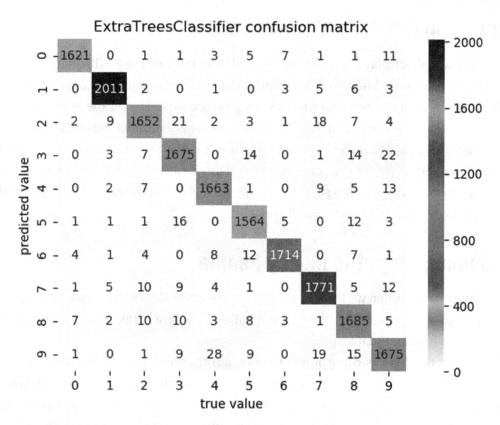

Figure 8-1. Confusion matrix

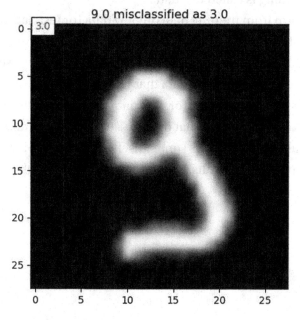

Figure 8-2. First incorrect prediction

Explaining MNIST to Money People

The code example displays how well we learned from training the MNIST data set. MNIST data represents images of digits from 0 to 9. Each element in the data set consists of the image as a set of pixels and what the image represents. That is, the set of pixels (like the picture we see on a television screen) represents a number between 0 and 9.

For instance, the first data element in the MNIST data consists of a set of pixels that represent the digit 0. So, we can train the entire data set based upon the knowledge that each data element consists of a set of pixels upon which a digit image is composed and the actual digit as a number between 0 and 9.

Explaining Output to Money People

The output shows us that we learned everything about the data set because training performance is 1.0 (or 100%). This sounds great, but we have to take into account how well we learn from new data.

Typically, machine learning practitioners slice off a piece of data from the data set and hide this from the training process. That is, they only train on a portion of the data and use what is learned on hidden (or untouched) data that was sliced off prior to training. The data that we learn from is called training data and the hidden data that is untouched during training is called test data.

Using what we learn from training data on new data (or test data) is critical because we predict future performance based upon test data. The reason we use test data for predicting the future is that the technical training process has *never* seen the test data. So, test data is representative of future data that we collect.

Training performance on test data (or new data) was 97.32%. So, training was successful because we are confident that over 97% of our predictions will be correct.

The rest of the output just displays some misclassifications (or errors) to help the machine learning expert verify results. Keep in mind that although performance is over 97%, we still have 2.68% error.

Explaining the Confusion Matrix to Money People

The confusion matrix got its name because it can be confusing how each of the numbers in the matrix is derived. However, it is really simple to explain the actual output from the matrix.

The numbers down the left-hand side represent predictions (labeled as predicted value) for each digit, and the numbers across the bottom represent actual values (labeled as true value) for each digit. Correct predictions are the numbers on the diagonal from top left to bottom right. Incorrect predictions are the numbers *not* on the diagonal.

Our confusion matrix shows that we predicted *1621* 0 digits correctly (top left). As we move down the diagonal, we see that we predicted *2011* 1 digits correctly and so on.

To see prediction performance in more detail, we can look at predictions for each digit or how actual digits were predicted. To analyze predictions for each digit, we look at row values. To analyze how actual digits were predicted, we look at column values.

Let's first look at digit 0 predictions, which are located along the top row from left to right. The first row value is 1621 (top left corner of confusion matrix), which represents correct predictions. So, we correctly predicted digit 0 1621 times. The remaining numbers along the row represent when we predicted digit 0, but the actual digit was some other value. For example, the last value in the top row is 11. So, we made eleven incorrect digit 0 predictions when the actual value was digit 9.

Now, let's look at predictions when the actual value was digit 0 that are located up and down the first column. The first column value is 1621. Since this number is on the diagonal, it represents correct predictions of digit 0. However, digit 0 was incorrectly predicted as digit 2 two times, digit 5 one time, digit 6 four times, digit 7 one time, digit 8 seven times, and digit 9 one time.

So, the results of a confusion matrix can be interpreted in two ways. One way is to look at prediction performance. The other way is to look at how an actual value was predicted. However, both ways lead to the same result.

For instance, a definite area for further study is to find out why digit 9 has so many incorrect predictions. The worst culprit is the value that shows twenty-eight (28) incorrect predictions. We can view this as predicting digit 9 incorrectly twenty-eight times when the actual value was digit 4. Conversely, we can view this as digit 4 being incorrectly predicted as digit 9 twenty-eight times.

The confusion matrix is an excellent way to see how well we predicted based on new data. It also provides a way to identify problem areas. For instance, we incorrectly predicted digit 9 more times than any other digit. So, machine learning experts can identify where more work needs to be done to increase performance.

Explaining Visualizations to Money People

The visualization shows the first incorrect prediction we made during training. We can clearly see that the actual digit is 9 (the large image in the middle), but it was classified as digit 3 (the small image in red positioned top left). Clues like this visualization can save time and money because we can at least speculate why training thought that the digit was 3 when it was really 9. Maybe training somehow saw the image with the top left opening closed, which would make it look like a 9.

Value and Cost

Digit recognition is a valuable launching point onto more complex image recognition ventures like face recognition. What we learn from image recognition contributes directly to face recognition learning exercises.

We demonstrated that our training provides over 97% accuracy with new data. That is, we know that at least 97 out of 100 times our predictions are correct. So, we can use what we learned on other machine learning ventures with confidence.

We were also able to assess the literal cost. Our training cost was 2.68%. That is, we made incorrect predictions less than 3 times out of 100.

Finally, we can easily identify areas for potential improvement and cost savings with the help of the confusion matrix and visualization. The confusion matrix showed us where training misfired. The visualization gives clues as to why training may have made incorrect predictions. Although we only showed one visualization, we can produce visualizations of all of the incorrect predictions for further study.

The ability to efficiently locate potential problem areas saves machine learning experts time and money because they at least have educated clues where to focus their energy. Without such clues, they would be experimenting in the dark.

Training exercises with image recognition also offer potential competitive advantage. We can use such exercises to feed more complex image recognition experiments like face and text recognition. Image recognition offers the fundamentals to artificially recognize what people are communicating and how they appear. Since there is an incredible rush to move into machine learning, we can learn from this example its potential benefits and costs.

Finally, image data like MNIST makes a strong case why machine learning experts need data in its raw form. Image data (of any kind) does not lend itself to storage in or access from relational or legacy systems. Machine learning experts need to get this data as it comes into the organization before others process and dilute it from its original form.

Computational expense is of course much harder to demonstrate to money people. However, this extremely simple example consumes over one minute just to run an already tuned algorithm. So, as we create more complex examples that are more computationally expensive, we can show money people more direct benefits like the ability to decipher written numbers accurately.

fetch_lfw_people Value and Cost

We continue with the fetch_lfw_people code example that was previously tuned, which is shown in Listing 8-2. We load the data, run the machine learning algorithm, and display results. We then explain results in business value and costs terms.

Listing 8-2. fetch_lfw_people value and cost

```
import numpy as np
from random import randint
from sklearn.decomposition import PCA
from sklearn.model_selection import train_test_split
from sklearn.svm import SVC
from sklearn.metrics import classification_report
import matplotlib.pyplot as plt
import seaborn as sns

def find_misses(test, pred):
    return [i for i, row in enumerate(test) if row != pred[i]]

def find_hit(n, ls):
    return True if n in ls else False

def build_fig(indx, pos, color, one, two):
    X_i = np.array(X_test[indx]).reshape(50, 37)
    t = targets[y_test[indx]]
    p = targets[y_pred[indx]]
```

```
    ax = fig.add_subplot(pos)
    image = ax.imshow(X_i,  cmap='bone')
    ax.set_axis_off()
    ax.set_title(t)
    ax.text(one, two, p, color=color,
            bbox=dict(facecolor='white'))

def chk_acc(rnds):
    logic = [1 if y_test[row] == y_pred[row] else 0
                for row in rnds]
    colors = ['g' if row == 1 else 'r' for row in logic]
    return colors

if __name__ == "__main__":
    br = '\n'
    X = np.load('data/X_faces.npy')
    y = np.load('data/y_faces.npy')
    bp = np.load('data/bp_face.npy')
    bp = bp.tolist()
    images = np.load('data/faces_images.npy')
    targets = np.load('data/faces_targets.npy')
    X_train, X_test, y_train, y_test = train_test_split(
        X, y, random_state=0)
    pca = PCA(n_components=0.95, whiten=True, random_state=0)
    pca.fit(X_train)
    X_train_pca = pca.transform(X_train)
    X_test_pca = pca.transform(X_test)
    svm = SVC(**bp)
    svm.fit(X_train_pca, y_train)
    y_pred = svm.predict(X_test_pca)
    print ()
    cr = classification_report(y_test, y_pred)
    print (cr)
    misses = find_misses(y_test, y_pred)
    miss = misses[0]
    hit = 1
```

```
X_hit = np.array(X_test[hit]).reshape(50, 37)
y_test_hit = targets[y_test[hit]]
y_pred_hit = targets[y_pred[hit]]
X_miss = np.array(X_test[miss]).reshape(50, 37)
y_test_miss = targets[y_test[miss]]
y_pred_miss = targets[y_pred[miss]]
fig = plt.figure('1st Hit and Miss')
fig.suptitle('Visualize 1st Hit and Miss',
             fontsize=18, fontweight='bold')
build_fig(hit, 121, 'g', 0.4, 1.9)
build_fig(miss, 122, 'r', 0.4, 1.9)
rnd_ints = [randint(0, y_test.shape[0]-1)
            for row in range(4)]
colors = chk_acc(rnd_ints)
fig = plt.figure('Four Random Predictions')
build_fig(rnd_ints[0], 221, colors[0], .9, 4.45)
build_fig(rnd_ints[1], 222, colors[1], .9, 4.45)
build_fig(rnd_ints[2], 223, colors[2], .9, 4.45)
build_fig(rnd_ints[3], 224, colors[3], .9, 4.45)
plt.tight_layout()
plt.show()
```

Your output from executing Listing 8-2 should resemble the following:

	precision	recall	f1-score	support
0	1.00	0.64	0.78	28
1	0.76	0.92	0.83	63
2	0.91	0.88	0.89	24
3	0.88	0.92	0.90	132
4	0.74	0.85	0.79	20
5	1.00	0.64	0.78	22
6	0.90	0.85	0.88	33
micro avg	0.86	0.86	0.86	322
macro avg	0.89	0.81	0.84	322
weighted avg	0.87	0.86	0.86	322

Listing 8-2 also displays Figures 8-3 and 8-4. Figure 8-3 shows a visualization of the first correct prediction and the first incorrect prediction. Figure 8-4 shows a visualization of four random predictions.

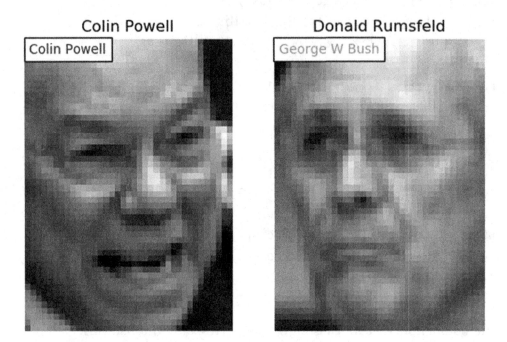

Figure 8-3. *First correct and first incorrect prediction*

Colin Powell

Colin Powell

George W Bush

George W Bush

Ariel Sharon

Ariel Sharon

Hugo Chavez

George W Bush

Figure 8-4. *Four random predictions*

Explaining fetch_lfw_people to Money People

The code example displays how well we learned from training the fetch_lfw_people data set. The fetch_lfw_people data set is a collection of JPEG images of famous people. Each element in the data set consists of the image and the person represented by the image.

Explaining Output to Money People

Training performance on test data (or new data) was 87%. This value is displayed at the bottom of the *precision* column on the left. So, we are confident that 87% of our predictions will be correct. But, this also means that we are incorrect 13% of the time. Of course, this example is a very simple and inexpensive one. As facial recognition technology continues to improve dramatically, learning accuracy approaches perfection. However, costs associated with capturing images, extracting samples to create templates, comparing collected data with existing templates, and matching collected data with templates at an industrial level can be expensive.

Although facial recognition is a very complex topic, the visualizations make it easy to see how well we learned from the data. We only displayed four random predictions, but we could have displayed many more for further analysis.

Explaining Visualizations to Money People

Both visualizations shown in Figures 8-3 and 8-4 represent predictions we made from what we learned from the data. Figure 8-3 shows a correct prediction (name in *green* embedded in the picture) and an incorrect prediction (name in *red* embedded in the picture). Figure 8-4 shows four random predictions. Notice that we made three correct predictions and only one incorrect one. That is, we correctly predicted Colin Powell, George W. Bush, and Ariel Sharon while we incorrectly predicted Hugo Chavez as George W. Bush.

Value and Cost

Facial recognition is the fastest growing biometric technology with the sole purpose of identifying human faces. Excellent facial recognition technology is already being used by Apple's iPhone X to unlock a smartphone.

A relevant area where facial recognition is critical is security. Organizations can protect their premises with this technology by tracking both employee and visitor movement in secure areas.

Another area is integration with existing software. Current facial recognition technology tools work well with existing security software. Such easy integration is great for business because organizations don't need to spend additional money and time redeveloping their own systems to work with facial recognition technology.

Current facial recognition technology accuracy is higher than ever before with the advent of 3D facial recognition technology and infrared cameras. Accuracy, of course, is extremely important because false identification can be detrimental.

Facial recognition systems can be fully automated. So, organizations won't need employees to monitor cameras.

Finally, time fraud can be drastically reduced. Since everyone must pass a face-scanning device to check in (or check out) for work or visit the premises, organizations don't have to worry about buddy favors from security staff members. Also, the process is much faster because technology controls the check-in or check-out process. Technology can also keep a log of activities if problems occur.

However, data capture and processing costs can be high depending on how much data needs to be collected and processed. Initial and ongoing technology costs can also be substantial. Don't forget that organizations still need competent people to monitor, fix, update, and upgrade the technology. The idea is to shift costs from manual processes to automated processes. In the short term, costs can be high, but automation should significantly reduce costs in the long term, improve process efficiency, and reduce human error.

Image quality can also be an issue. For instance, organizations can store images with different levels of quality. High-quality images require more storage space, but are better when matching against an actual face in real time. Of course, more storage is costly.

Surveillance angle is also an issue. Capturing a new image must be processed at different angles because an actual face in real time can be seen at different angles. People can also appear different by wearing sunglasses, being unshaven, emanating different facial expressions, and even having a different haircut.

The best technology takes these issues into account. Of course, the best technology costs money. However, people can't do what computers can. One example is simultaneous comparison of faces against a database of thousands.

As facial recognition options become more cost competitive, organizations may have no choice but to get involved with the technology. We have already discussed the value and costs of facial recognition. So, cost savings with automation, higher accuracy, and less fraud offer competitive advantages that cannot be ignored.

fetch_20newsgroups Value and Cost

We finish with the fetch_20newsgroups code example that was previously tuned, which is shown in Listing 8-3. We adopt a more realistic scenario by removing headers, footers, and quotes from the subset of documents that we train upon.

The reason we do this is to remove clues as to the subject of the document to ensure that machine learning algorithms can correctly identify meaning from text that is not easily identifiable. Keep in mind that headers, footers, and quotes are not typically in most written documents and written communications.

We load the data, run the machine learning algorithm, and display results. We then explain results in business value and costs terms.

Listing 8-3. fetch_20newsgroups value and cost

```
import numpy as np
from sklearn.datasets import fetch_20newsgroups
from sklearn.feature_extraction.text import TfidfVectorizer
from sklearn.naive_bayes import MultinomialNB
from sklearn.pipeline import make_pipeline
from sklearn.metrics import f1_score, confusion_matrix
import matplotlib.pyplot as plt
import seaborn as sns

def predict_category(s, m, t):
    pred = m.predict([s])
    return t[pred[0]]

if __name__ == "__main__":
    br = '\n'
    train = fetch_20newsgroups(subset='train')
    test = fetch_20newsgroups(subset='test')
    categories = ['rec.autos', 'rec.motorcycles', 'sci.space', 'sci.med']
    train = fetch_20newsgroups(subset='train', categories=categories,
                               remove=('headers', 'footers', 'quotes'))
    test = fetch_20newsgroups(subset='test', categories=categories,
                              remove=('headers', 'footers', 'quotes'))
    targets = train.target_names
    print ('targets:')
    print (targets, br)
    bp = np.load('data/bp_news.npy')
    bp = bp.tolist()
```

```
print ('best parameters:')
print (bp, br)
mnb = MultinomialNB(alpha=0.01, fit_prior=False)
tf = TfidfVectorizer(ngram_range=(1, 1), use_idf=False)
pipe = make_pipeline(tf, mnb)
pipe.fit(train.data, train.target)
labels = pipe.predict(test.data)
f1 = f1_score(test.target, labels, average='micro')
print ('f1_score', f1, br)
labels = pipe.predict(test.data)
cm = confusion_matrix(test.target, labels)
plt.figure('confusion matrix')
sns.heatmap(cm.T, square=True, annot=True, fmt='d', xticklabels=train.
            target_names, yticklabels=train.target_names, cbar=False)
plt.xlabel('true label')
plt.ylabel('predicted label')
plt.tight_layout()
print ('***PREDICTIONS***:')
doc1 = 'imagine the stars ...'
doc2 = 'crashed on highway without seatbelt'
doc3 = 'dad hated the medicine ...'
y_pred = predict_category(doc1, pipe, targets)
print (y_pred)
y_pred = predict_category(doc2, pipe, targets)
print (y_pred)
y_pred = predict_category(doc3, pipe, targets)
print (y_pred)
plt.show()
```

Your output from executing Listing 8-3 should resemble the following:

```
targets:
['rec.autos', 'rec.motorcycles', 'sci.med', 'sci.space']

best parameters:
{'tfidfvectorizer__use_idf': False, 'tfidfvectorizer__ngram_range': (1, 1),
'multinomialnb__fit_prior': False, 'multinomialnb__alpha': 0.01}

f1_score 0.8680555555555556

***PREDICTIONS***:
sci.space
rec.autos
sci.med
```

Listing 8-3 also displays Figure 8-5. Figure 8-5 shows the confusion matrix.

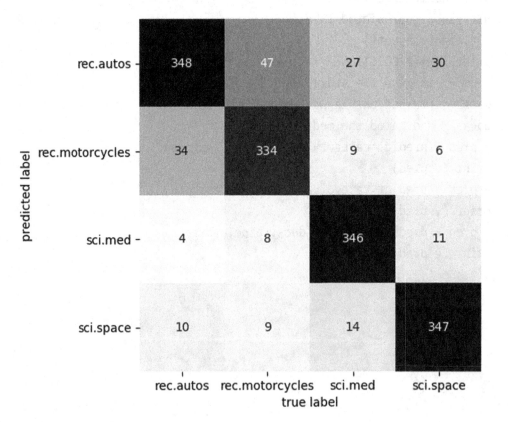

Figure 8-5. *Confusion matrix*

Explaining fetch_20newsgroups to Money People

The code example demonstrates how well we learned from training the fetch_20newsgroups data set. The fetch_20newsgroups data set is a collection of 18000 newsgroup posts on 20 topics. For this training example, we filtered down the data set into four subsets – autos, motorcycles, space, and medicine – to simplify and reduce computational expense.

Explaining Output to Money People

We first display the subsets upon which we learn from. Next, we display the best parameters so that the machine learning expert can appropriately build the algorithm that provides the best performance.

Training performance on test data (or new data) was almost 87%. This value is displayed as the f1_score. So, we are confident that almost 87% of our predictions will be correct, but this also means that we are incorrect a bit over 13% of the time.

Finally, we make predictions from three simple documents. Notice that we as humans can easily predict what category *imagine the stars...* belongs to, but the learning experiment doesn't have the vast knowledge about words that we do. It works with complex document manipulation techniques and machine learning algorithms to produce pretty good results.

Explaining the Confusion Matrix to Money People

Since we already explained what a confusion matrix is and how it can be analyzed in the first code example, we won't repeat it here. However, it does exhibit some interesting insights.

Notice on the first row, second column that we made 47 incorrect predictions. In this case, we incorrectly predicted autos when the actual values were motorcycles. This makes some sense because the learning algorithms have more trouble distinguishing between two types of vehicles than other incorrect predictions.

Conversely, we incorrectly predicted motorcycles incorrectly 34 times when the actual values were autos. We see again that the learning algorithms had trouble distinguishing between the two types of vehicles, so the majority of incorrect predictions were made when trying to distinguish between two types of vehicles.

Value and Cost

Extracting meaning from text is typically referred to as text mining. Text mining is a means to capture high-quality information from text. High-quality information has no value unless it leads to business insights (or value).

Since most data collected by organizations is never analyzed, text mining offers an efficient and hopefully effective means to do so. Just because an organization efficiently collects and mines text doesn't mean that it can create value.

Value can be created in at least three impactful areas. First, it can enhance compliance and threat detection. Second, it can foster customer engagement. Third, it can facilitate better decision-making.

Mining text can greatly impact compliance issues by providing early fraud detection such as money laundering. Organizations also need ways to automate compliance to policies and procedures. Text mining can automate such processes by detecting noncompliance from textual inputs such as online forms, e-mails, texts, and other messaging services. Threats to security can also be detected by mining messages flowing in and out of an organization.

Costs can definitely be reduced through automation of compliance and threat detection services because less people are needed to manage administration of such procedures. Human error is also reduced if less people are involved.

Interaction with customers offers incredible opportunities for text mining. For instance, Amazon profiles user preferences to better inform customers of products they may want to purchase.

Text mining is a natural vehicle to get an idea of what customers are thinking. Maybe customers are dissatisfied with current services. Maybe customers want a product or service that is not currently offered. Once an organization figures out how to gain insight from customers, it can automate such processes. Automation saves time and money and reduces human error.

If an organization can gain insight into customer thinking, value is created. If an organization can automate such insights, competitive advantage is created. Imagine if your competitors aren't involved in text mining. Even if they are involved with text mining, your organization's competitive advantage is the ability to more efficiently and effectively manage customer insights.

Finally, text mining can lead to better business decisions. Data analysts and managers need data to provide business with accurate insights. Text mining offers a powerful vehicle for gaining accurate insights.

Of course, initiating, implementing, administrating, and monitoring text mining activities come with costs. First, you need competent text mining people, sufficient computing resources, top-level support for leveraging text mining for business insight creation, and the ability to form teams composed of a mixture of technical and business members.

Text mining experts are great at implementing algorithms, but need help in defining the appropriate business insights an organization is seeking. So, a diverse team populated with a variety of team members with different aspects of the organization is very much needed. Competitive advantage can only be gained if your organization is better at text mining than your competition.

Index

© David Paper 2020
D. Paper, *Hands-on Scikit-Learn for Machine Learning Applications*,
https://doi.org/10.1007/978-1-4842-5373-1

Printed in the United States
By Bookmasters